Edward C. (Edward Charles) Pickering

An investigation in stellar photography

conducted at the Harvard college observatory with the aid of an

appropriation from the Bache fund

Edward C. (Edward Charles) Pickering

An investigation in stellar photography
conducted at the Harvard college observatory with the aid of an appropriation from
the Bache fund

ISBN/EAN: 9783337377458

Printed in Europe, USA, Canada, Australia, Japan

Cover: Foto ©berggeist007 / pixelio.de

More available books at **www.hansebooks.com**

AN INVESTIGATION

IN

STELLAR PHOTOGRAPHY

CONDUCTED AT

THE HARVARD COLLEGE OBSERVATORY

WITH THE AID OF AN APPROPRIATION FROM THE BACHE FUND

BY

EDWARD C. PICKERING

EXTRACTED FROM VOLUME XI. OF THE MEMOIRS OF THE AMERICAN ACADEMY

CAMBRIDGE
JOHN WILSON AND SON
University Press
1886

[Investigations on Light and Heat published with Appropriation from the Rumford Fund.]

IV.

Stellar Photography.

By EDWARD C. PICKERING.

Presented March 10, 1886.

THE experiments to be described below were mainly conducted by the aid of an appropriation made in June, 1885, from the Bache Fund, of the National Academy of Sciences. Numerous preliminary experiments had been made with a grant from the Rumford Fund of the American Academy. My attention was directed to stellar photography in 1882, by Mr. W. H. Pickering. Many of the preliminary experiments were made by him, and his advice was followed regarding the photographic processes to be employed. In the later work, he has rendered important aid by his advice, and by making many auxiliary experiments in his photographic laboratory at the Massachusetts Institute of Technology. The actual exposures and development of the photographs have been made by several assistants in turn. From June 11, 1885, to October 16, 1885, this work was intrusted to Mr. A. H. Whittemore; from October 17, 1885, to January 26, 1886, to Mr. H. Helm Clayton; and from February 2, 1886, to the present time, to Mr. Willard P. Gerrish. To the skill and perseverance shown by these gentlemen the success attained is largely due.

The following subjects are discussed below in turn: —

History of Stellar Photography.
Preliminary Experiments, 1882 to 1885.
Description of the Photographic Apparatus finally adopted.
Theoretical Considerations entering into the Problem of Stellar Photography.
Trails formed by Stars when their apparent Motion is not wholly corrected
    by the Motion of the Telescope.
Construction of Charts by Photography.
Stellar Spectra.
Brighter Stars in the Pleiades.
Close Polar Stars.

HISTORY.

Stellar photography originated in an experiment made at the Harvard College Observatory on July 17, 1850. Under the direction of Professor W. C. Bond, Mr. J. A. Whipple placed a sensitive daguerreotype plate in the focus of the fifteen-inch equatorial, which by means of its driving-clock was kept pointed upon the star a *Lyræ*. A satisfactory image of the star was thus obtained. Subsequently, the double star a *Geminorum* gave an elongated image, evidently due to its two components. Objects as bright as these gave but faint images, and no impression was obtained from the Pole-star, however long the exposure continued. The experiment was repeated with various stars and clusters, but the work was finally abandoned owing to the imperfections of the driving-clock and the lack of sensitiveness of the plates (Harvard Observatory Annals, I. cxlix, clvii, clxv). Both of these difficulties were partially remedied in 1857, and the research was resumed by Professor G. P. Bond. The driving-clock, regulated by a conical pendulum, was replaced by a larger clock, controlled by a Bond's spring-governor. The introduction of the collodion process had greatly reduced the photographic difficulties, and furnished plates of much greater sensitiveness. The results of this investigation are contained in three important papers, which were published in the Astronomische Nachrichten, XLVII. 1, XLVIII. 1, and XLIX. 81, and have now become classical. The first of these papers states that on April 27, 1857, an impression of the double star ζ *Ursæ Majoris* was obtained in eight seconds. An exposure of two or three seconds was afterwards found to be sufficient to produce an impression of the brighter star; and when the telescope was at rest, a trail was obtained as the image of the star a *Lyræ* traversed the plate. A series of measures was made of the position angle and distance of the companion of ζ *Ursæ Majoris*. The probable error of a single photographic distance was found to be ±0".12. The faintest star photographed was the companion of ε *Lyræ*, which has a magnitude of 6.0. The second paper contains a careful study of the distance of the components of ζ *Ursæ Majoris*. The result was 14".21 ± 0".013 from measures of sixty-two images taken on eight nights. The extreme variations in the results of these eight nights was only 0".08, and the probable error of a single measure was ±0".05. The third paper is devoted to a discussion of the relative brightness of the stars as indicated by the diameters of the photographic images. The advantages of photography as a means of locating the stars in a cluster are clearly stated. In fact, nearly all the arguments now offered in favor of photography will be found in these admirable papers.

These experiments were soon after repeated by Mr. De la Rue and by Mr. Rutherfurd. A much more extended investigation was undertaken in 1864 by Mr. Rutherfurd, and continued by him during many years. A brief notice of this work was published in the American Journal, XXXIX., 1865, 304. One of his photographs of the *Pleiades* was measured and discussed by Dr. B. A. Gould, in 1866, in the Astronomische Nachrichten, LXVIII. 184. A list of the clusters photographed by Mr. Rutherfurd is given by Professor Holden, in the Smithsonian Miscellaneous Collections, 311, page 89. The faintest stars shown in these photographs are probably not far from the ninth magnitude. By the continued improvement in photographic processes, each experimenter after a few years has a great advantage over his predecessors. The invention of dry plates simplified the work of taking a photograph, permitted an indefinite prolongation of the time of exposure, and ultimately greatly increased the sensitiveness of the film. Aided by these advantages, Dr. Henry Draper attacked the problem with his usual skill and perseverance. On March 11, 1881, he obtained a photograph of the Nebula of Orion, in which a star is shown whose photometric magnitude is about 14.7. This star is barely visible with a telescope of the same aperture as that with which the photograph was taken. The photographic plate, accordingly, had now become as efficient an instrument of research as the eye itself, by means of its power of accumulating the energy radiated upon it. For a further discussion of the faintest stars thus photographed, see Washington Observations, 1878, page 226, and Proceedings of the American Academy, XX. 407. But for Dr. Draper's untimely death, he would doubtless have been the first to accomplish the striking experiment of photographing a star too faint to be seen in the largest telescopes. This result was apparently soon after attained by Mr. A. A. Common, in his beautiful photograph of the Nebula of Orion. It is not safe to draw conclusions from a single star, on account of the effect of color. A red and a blue star may produce photographic images of equal intensity, although to the eye their brightness may differ by several magnitudes. A comparison of the photograph of Mr. Common with the catalogue of Bond (Annals Harvard College Observatory, V. 270) has been made by the writer (Proc. Amer. Acad., XX. 407). From this it appears that the number of stars contained in the photograph and not given in the catalogue is about equal to the number of stars in the catalogue which are wanting in the photograph. We may therefore conclude that the limiting magnitude for the photograph does not differ greatly from that of the faintest stars visible to the eye. The largest telescopes add but few stars to those given by Bond, although this nebula has been thoroughly scrutinized with the reflectors of the

Earl of Rosse and the Washington refractor. An important part of the work of Dr. Gould at the Cordoba Observatory consisted in securing photographs of the principal clusters in the southern hemisphere, but the results have not yet been published.

During the past three or four years, stellar photography has been pursued at various observatories. Dr. Gill has undertaken to make a map of the southern heavens by photography. A catalogue of five hundred stars, whose light was determined from their photographs, was published by the Rev. T. E. Espin in the Proceedings of the Liverpool Astronomical Society, Transactions, No. 3, 1884. The most elaborate investigation is that of the MM. Henry, of the Paris Observatory. Admirable photographic maps of the stars have been constructed with the intention of substituting them for the charts of Peters and Chacornac. It is impossible, however, to make a statement of the photographic work in progress at the present time, on account of the rapid advances now being made in this department of astronomy.

In 1863, Dr. Huggins obtained a photographic image of the spectrum of Sirius, but it was so ill-defined that it presented no indications of lines. (Phil. Trans., 1864, p. 428.) The first successful photograph of the spectrum of a star was obtained by Dr. Henry Draper, in 1872. A spectrum of Vega was taken, in which four lines were visible. (Amer. Jour. Sci., 1879, XVIII.) The work was afterwards resumed by each of these astronomers, but was confined to the brightest stars. (Proc. Amer. Acad., XIX. 231; Phil. Trans., 1880, p. 669.) The method employed was the same in both cases, and consisted in concentrating the light of the star by means of a large telescope upon the slit of a spectroscope placed in its focus. A narrow slit was necessary to secure good definition, and very perfect adjustment of the telescope was required to keep the image of the star upon the slit.

PRELIMINARY EXPERIMENTS.

A great variety of preliminary experiments were made in connection with this research, by the aid of an appropriation from the Rumford Fund. In 1882, a photographic lens having an aperture of 7 inches and a focal length of 37 inches was procured, and attached to the equatorial mounting in the west dome of the Harvard College Observatory. Afterwards the camera containing this lens was mounted in the meridian and directed towards the equator. The plate-holder was attached to a car, which was drawn by clockwork at the rate of 0.16 in. per minute from west to east. If the plate was at rest, each star in turn, as it approached the meridian, would form an image on the western edge of the plate, gradually traversing it along a horizontal line. The velocity would be proportional to the focal length, and in this case would be 0.16 in. per minute. If then the plate was moving with the same velocity in the same direction, the star would evidently form a circular dot upon it. A panoramic photograph was thus taken, and the region covered was only limited by the length of the plate. It was proposed to expose in this way a series of plates, one following the other, and thus photograph zones several hours in length. On December 6, 1882, the first photograph was taken by this method. On one plate 462 stars were counted where only 55 occur in the Uranometria Argentina. The faintest stars were probably of about the ninth magnitude. The images were large and distorted, owing to defects of the lens and to imperfect adjustment of the apparatus. Otherwise, much fainter stars would have been photographed. For zones in other declinations a conical motion could be given to the car by means of a series of curved tracks, or by other devices. On February 21, 1883, a photograph of Orion was obtained by Mr. W. H. Pickering with a small Voigtländer camera and without clockwork. Although the aperture was only 1.6 in. and the focus 5.2 in., trails were obtained of stars as faint as the fifth magnitude. Still better results were obtained with a Voigtländer No. 4 lens (Series B), having an aperture of 2 inches and a focal length of 7 inches.

Attaching clockwork and giving an exposure of thirty minutes, a photograph was obtained of all the stars down to the eighth magnitude in a region about 15° square. Plans were prepared for photographing a large part of the sky in this way. The lens was mounted equatorially with a long rod attached to the declination axis. A ribbon of brass was fastened to the end of the rod, and by clockwork was wound uniformly around a drum. A second series of photographs was undertaken with this same instrument. Six portions of the plate could be brought in turn to the focus of the lens, and three exposures were to be made on each part of the plate. No clockwork was to be used, and the different regions taken on the same part of the plate were to be distinguished by the length of the trails. This was accomplished by giving to the various regions a single exposure, a long exposure followed by a short one, or a short exposure followed by a long one. In order to direct the telescope quickly to the desired portion of the sky, notches were made in both declination and right ascension circles at intervals of 15°. To reduce the results to a scale of stellar magnitudes, trails were taken with a series of apertures, whose area formed a measure of the light transmitted. Another method which is also applicable when clockwork is used, consists in attaching a prism of very small angle to the centre of the object-glass. The portion of the light falling upon this will form a companion to each star. The ratio of the true brightness of the star to that of its companion is a constant quantity, and is readily determined from the absorption of the prism and the ratio of its area to that of the uncovered portion of the lens. If the stars and companions are then measured upon any empirical scale, this ratio of light serves to reduce the results to absolute measures.

To remedy the variation arising from the color of the stars, an attempt was made to render the light monochromatic. One method proposed for this purpose consisted in using as an object-glass an uncorrected lens. The focus of this varies very greatly with the color of the light. The rays out of focus would be spread over so large a circle that they would have but little effect upon the image. By covering the centre of the lens by a diaphragm, or by a prism of small angle, the light at the centre of the image of the star would be nearly monochromatic. By taking a series of plates with various foci, the relative intensity of the light of each different color could be determined. The most important of these methods, and the results attained by them, were described and exhibited to the Royal Astronomical Society at their meeting on June 8, 1883. (Observatory, VII. 199; Astron. Register, XXI. 149.)

PHOTOGRAPHIC APPARATUS.

A second series of experiments was undertaken in March, 1885, with results that seemed to justify their repetition on a larger scale. An application was accordingly made to the Board of Directors of the Bache Fund, and an appropriation was granted by means of which the following work was accomplished. A Voigtländer lens having an aperture of 8 inches and a focal length of about 45 inches was obtained, and intrusted to Messrs. Alvan Clark and Sons for correction and mounting. Errors which are quite inappreciable in ordinary photographic work may be large enough to ruin a stellar photograph. This was found to be the case with the lens mentioned above. It was also desirable that its focal length should be 114.6 cm., so that the scale of the photographs should be as nearly as possible 2 cm. to 1°, which is the scale of the maps of the Durchmusterung. These difficulties were finally overcome, but it was necessary to regrind two of the glass surfaces. The lens was next mounted equatorially, as is shown in Plate I. The brass tube carrying the lens is screwed into the end of a steel tube which is mounted in trunnions at the ends of a large fork. This fork forms the prolongation of the polar axis, and possesses some important advantages over the usual equatorial mounting. It is much lighter than the German form in common use, since the tube is not eccentric, and therefore no counterpoise is required. It does not have to be reversed when a star crosses the meridian, and in fact any star may be followed uninterruptedly from rising to setting. The advantage over the English form of mounting is, that polar stars can be observed without difficulty. The objection to its general introduction is, that the end of the polar axis would interfere with observations made near the pole. A declination circle is attached to the tube and divided into single degrees. A closer division is not necessary, since a small error in setting would only affect the part of the plate on which a star would impress itself. Notches are cut in the edges of the circle for each degree, every tenth notch being made deeper than the others. A catch at the end of a steel spring is attached to the fork, and enables the telescope to be moved quickly and with precision either 1° or 10° at a time, as may be desired. The edges of the catch are ground at a greater angle than that of friction, and thus permit the telescope to be moved when sufficient pressure is applied. The right ascension circle consists of an iron wheel 67 cm. in diameter, in the edges of which 720 teeth are cut with great care. A screw turns in these teeth, forming a worm and wheel. One tooth corresponds to half a degree, or to two minutes of time in right ascension. An index serves to set the telescope within a minute or less, and this is close enough for the present purpose. The screw is driven by clockwork controlled by a Bond spring-governor, the weights being connected by the Huyghenian arrangement, so that they may be rewound without stopping the clock-work. It is difficult to make a pendulum keep accurate time in a location where the variations in temperature are great. Moreover, to photograph the fainter stars it is necessary that the clockwork should run for a long time with a high degree of accuracy. These conditions are fulfilled by controlling the Bond spring-governor electrically by a clock mounted in the clock-room of the Observatory. If the polar axis is parallel to the axis of the earth, the telescope can thus be made to follow a star for an indefinite time, as closely as the clock can be made to run. The error from this source is therefore entirely inappreciable. The differential refraction of the air would be a more serious source of error than this. The electrical contacts occurred in the clock every two seconds, and that corresponding to the fifty-eighth second was omitted. A spring-governor will run more satisfactorily when the contact occurs every second, or oftener; but with the present arrangement the error could seldom exceed a fraction of a second. The fifty-eighth second was omitted by a secondary connection in the clock occurring once a minute, and not, as is sometimes done, by the omission of one tooth in a contact wheel. The removal of the secondary connection gave the signals uninterruptedly. Finally, the clock was arranged as a make-circuit and the governor required a break-circuit. This difficulty was remedied by throwing the clock out of the circuit and using it as a shunt to the battery. The circuit was thus closed both through the clock and the governor. The clock resistance was so small that but little of the current ordinarily passed through the governor. Every two seconds the clock circuit was broken, and the entire current passed through the governor. An incidental advantage of this arrangement is, that, as the circuit is never entirely broken, no spark occurs within the clock.

The frame carrying the polar axis rested on a bed-plate with adjusting screws so that it could be raised or lowered at either corner or moved in azimuth. The whole rested on a large stone imbedded in the ground and covered by a transit shed, which had been constructed for observing the Transit of Mercury in 1878. All the photographs were taken near the meridian, where the atmospheric absorption and refraction are least and vary most slowly. The stone did not prove to be sufficiently immovable when subjected to frosts and thaws, and consequently a circular level was attached to the frame of the instrument. The polar axis could thus always be brought back to the same altitude, whatever the motion of the stone.

The transit shed covering the instrument originally had a slit only one foot in width. This was widened to two feet and closed by two light shutters covered with canvas. It was expected that this would permit stars on the equator to be reached when one hour east or west of the meridian. When the instrument was mounted, the axis was found to be lower than had been anticipated, and the edges of the slit began to interfere when the telescope was no more than twenty minutes from the meridian. This difficulty was remedied in March, 1886, when the instrument was remounted. It was then placed on the pier used during the Transit of Mercury, which proved a much steadier support. It also permitted exposures of over an hour on the equator, and longer exposures near the pole, without interfering with the edges of the shutter.

All the photographs were taken on bromo-gelatine dry plates, having the dimension of eight inches by ten. They accordingly covered a region of 12° in declination and 10° in right ascension. On the equator this represents forty minutes of time. Most of the plates were made by Messrs. Allen and Rowell, of Boston, and were of the form known as the "Extra Quick." Cramer's developer was used, containing pyrogallic acid and carbonate of soda.

The photographic work may be divided into three classes. First, the telescope being at rest, photographing the trails of the stars; secondly, with the aid of clockwork, forming charts; and, thirdly, photographing the spectra of the stars. These subjects will be considered in three separate sections.

THEORETICAL CONSIDERATIONS.

The first case to be considered is that of a luminous point at rest. The formation of a chart of the stars when the instrument is perfectly adjusted is an example of this case. Secondly, let the luminous object be a line instead of a point. A moving luminous point, as a star when the telescope is at rest or the adjustment is imperfect, will be considered in this connection. The third case is that of a luminous surface, as a nebula or comet.

To compare the advantages of different forms of instrument, a discussion is given below to show the relation between the dimensions of the lens employed and the light of the faintest star that can be photographed with it.

The following notation will be used : —

a = aperture of lens.

f = focal length of lens.

t = proportion of light transmitted by lens.

d = diameter of the photographic image of the faintest star capable of impressing itself upon the plate.

m = brightness of such a star expressed in stellar magnitudes.

l = ratio of the light emitted by such a star to that of a star of magnitude zero.

T = time of exposure.

s = sensitiveness of plate measured by the amount of light required to produce the faintest perceptible photographic impression.

The value of l will be proportional to d^2 and to s, inversely proportional to a^2 and to t, and may be assumed to be inversely proportional to T. This last assumption is based on experiment, and seems to be justified within wide limits. Then $l = A \frac{d^2 s}{a^2 t T}$, in which A is a constant dependent on the units employed to indicate the magnitudes of the various quantities involved in this equation. Upon Pogson's system of magnitudes, $m = -2.5 \log l$, and hence

$$m = -2.5 (\log A + 2 \log d + \log s - 2 \log a - \log t - \log T).$$

The uncertainty in the value of d renders it difficult to compare results in different cases. The principal causes affecting it are, first, variations in the atmospheric refraction. With a given aperture, these variations may be regarded as causing a constant angular deviation of different portions of the beam; d will therefore be proportional to f. It will also increase rapidly if a increases. Secondly, spherical and chromatic aberration of the lens. This will have the same angular value in equally perfect lenses, so that when $\frac{a}{f}$ is constant it will be proportional to f. Its angular value will increase rapidly as a increases. Thirdly, diffraction from the edges of the lens. This error is ordinarily small, and, unlike the other sources of error, increases as the aperture diminishes. Fourthly, a chemical action occurs by which a decomposition of the salts of silver, at the point exposed to the light, extends to the adjacent particles. This effect is very marked in the case of the bright stars, and is of course independent of the dimensions of the lens. The increased size of the images of the bright stars is in part only due to this chemical action. It is also caused by the light diffused by the other sources of error mentioned above, and by the light reflected from the back of the plate. In the faint stars this chemical action is inappreciable, but becomes perceptible when the light is intense. It is therefore very difficult to compare two lenses theoretically, even if we are sure that they differ only in their dimensions. If we assume that d is a constant, so that the fourth of the above causes is the principal one to act, l will depend only on the aperture of the object-glass. Since one magnitude corresponds to a ratio of light of 2.512, an increase of the aperture of 1.585 should permit stars one magnitude fainter to be photographed in the same time. In like manner, increasing the aperture ten times should extend this limit by five magnitudes. In reality this is far from being the case, the actual increase in aperture required being much greater. Again, if the angular diameter of the images is nearly constant, d will be proportional to f, and l will remain the same as long as the angular aperture $\frac{a}{f}$ is constant. On this hypothesis similar lenses, whether large or small, could photograph equally faint stars. The fact lies somewhere between the two, and perhaps it would not be far from the truth to assume that the limiting amount of light was proportional to $\frac{a}{\sqrt{f}}$. When the positions of the stars are to be measured, a large telescope has a very great advantage. The scale is proportional to the focal length, and, since the errors of measurement are nearly constant when expressed linearly, this effect will be inversely as the focal length.

The second case to be considered is that in which the image of the star is slowly traversing the plate. Let r equal the velocity, or distance traversed per second. Then the time of exposure will equal that required by the star to move a distance equal to its diameter, and $T = \frac{d}{v}$. We therefore shall have $l = \frac{A \, ds \, v}{d^2 \, t}$, and as long as the instrument remains unchanged l is proportional to v. If the telescope is at rest and the motion of the star is due to the rotation of the earth, $v = \frac{2 \pi f}{86400} \cos \delta$, and $l = \frac{2 \pi f A \, ds \cos \delta}{86400 \, d^2 \, t}$, denoting the declination of the star by δ. When stars of different declinations are photographed with the same instrument, δ is the only variable in the right-hand member of this equation; hence we may write $l = C \cos \delta$, and l is proportional to the cosine of the declination of the star. We then have $m = -2.5 \log l = -2.5 \log C - 2.5 \log \cos \delta$. When we wish to compare the relative brightness of different stars from the intensities of the trails they leave, this correction must first be applied. The trails must first be reduced to a scale of stellar magnitudes, as will be described later (page 211), and then the correction 2.5 log cos δ added to each. This correction is facilitated by the aid of Table I., which gives for every degree of declination the correction in hundredths of a magnitude.

TABLE I.

δ	Magn.	δ	Magn.	δ	Magn.	δ	Magn.	δ	Magn.	δ	Magn.	δ	Magn.	δ	Magn.	δ	Magn.
0	0.00	10	0.02	20	0.07	30	0.16	40	0.29	50	0.48	60	0.75	70	1.16	80	1.90
1	0.00	11	0.02	21	0.07	31	0.17	41	0.31	51	0.50	61	0.79	71	1.22	81	2.01
2	0.00	12	0.02	22	0.08	32	0.18	42	0.32	52	0.53	62	0.82	72	1.28	82	2.14
3	0.00	13	0.03	23	0.09	33	0.19	43	0.34	53	0.55	63	0.86	73	1.34	83	2.29
4	0.00	14	0.03	24	0.10	34	0.20	44	0.36	54	0.58	64	0.90	74	1.40	84	2.45
5	0.00	15	0.04	25	0.11	35	0.22	45	0.38	55	0.60	65	0.94	75	1.47	85	2.65
6	0.01	16	0.04	26	0.12	36	0.23	46	0.40	56	0.63	66	0.98	76	1.54	86	2.89
7	0.01	17	0.05	27	0.13	37	0.24	47	0.42	57	0.66	67	1.02	77	1.62	87	3.20
8	0.01	18	0.05	28	0.14	38	0.26	48	0.44	58	0.69	68	1.07	78	1.71	88	3.64
9	0.01	19	0.06	29	0.15	39	0.27	49	0.46	59	0.72	69	1.11	79	1.80	89	4.40

It is generally sufficiently precise to carry the computations to tenths of a magnitude. In this case, Table II. is more convenient, especially for polar stars where the correction changes rapidly. The limits of declination within which the corrections of each tenth of a magnitude should be applied are given in this table. The correction for any declination given in the table is found by adding the whole number of magnitudes taken from the top of the column to the tenth of a magnitude at the beginning of the line. The correction for intermediate declinations is the same as that of the next smaller declination given in the table. Thus, the correction will be 0.0 for stars at any declination from 0° 0′ to 17° 15′ inclusive. It will be 1.0 for stars between 65° 22′ and 67° 30′.

TABLE II.

	0		1		2		3		4	
	°	′	°	′	°	′	°	′	°	′
.0	0	0	65	22	80	27	86	13	88	30
.1	17	16	67	40	81	18		33		38
.2	29	26	69	43	82	4		51		45
.3	37	25	71	34		47	87	8		52
.4	43	35	73	15	83	25		23		58
.5	48	39	74	46	84	0		37	89	3
.6	52	57	76	8		32		50		8
.7	56	40	77	22	85	1	88	1		13
.8	59	56	78	30		27		12		17
.9	62	49	79	31		51		21		21

When a star is very near the pole, the cosine of the declination will be proportional to the polar distance p. If this quantity is expressed in minutes of arc, the rate of motion will be proportional to $\frac{p}{3438}$. The correction in magnitudes will then be $2.5 \log .3438 - 2.5 \log p$, or $8.84 - 2.5 \log p$. The correction for the polar distances 1′, 10′, 100′, and 1,000′ will by this formula be 8.84, 6.34, 3.84, and 1.34. A direct computation from cos δ gives 8.84, 6.34, 3.84, and 1.36.

Owing to precession, the relative brightness of the trails of the stars will vary from year to year. The effect will of course be very small, except for stars near the pole. As shown above, the correction for declination is $2.5 \log \cos \delta$. The differential coefficient of this expression with respect to δ is $2.5 \times .434 \tan \delta$, or $1.085 \tan \delta$. If the precession in declination for n years is $\frac{20''.45 \, n \cos a}{206265''}$, the change in the correction is $.000107 \, n \cos a \tan \delta$.

A table of $133.7 \sin a \tan \delta$ is given in Oeltzen (Vol. I. p. xiii) for various values of a and δ. This table may be employed for the present purpose by placing the expression just obtained in the form $.0000008003 \, (133.7 \sin a \tan \delta) \, n$, and changing a by 6 hours. The sign is determined by the rule that for stars between 6ʰ and 18ʰ of right ascension the correction to the photographic brightness diminishes.

For stars in the immediate vicinity of the pole, when the polar distance expressed in minutes is p, $\tan \delta = \frac{3438}{p}$, and the correction for precession $0.37 \, \frac{n \cos a}{p}$.

All of these formulas are of course approximate, and, if used for intervals of time in which the change in right ascension is very great, the neglected terms may have to be considered.

The relative brightness of the trails of the brighter stars, and also of some faint close polars, is given in Table III. The successive columns give the number from the Harvard Photometry, the usual designation of the star, its right ascension and declination for 1885, its brightness to the eye, and the corresponding brightness of the photographic trail. The first eight magnitudes in the last column but one were taken from the Harvard Observatory Annals, XIV. 406. The others are taken from the Proceedings of the American Association, XXXIII. 1, except β *Ursæ Minoris*, which is taken from the Harvard Annals, XIV. 200, and σ *Octantis*, which is taken from the Uranometria Argentina, p. 131.

TABLE III.

H. P.	Desig.	R. A. 1885.	Dec. 1885.	Magn.	Trail.
		m. *s.*	*°* *'*		
1275	α Canis Majoris	6 40.1	−16 34	−1.4	−1.4
—	α Argus	6 21.4	−52 38	−0.8	−0.3
—	α Centauri . . .	14 31.8	−60 22	−0.1	−0.9
2400	α Bootis	14 10.4	+19 47	0.1	0.0
932	α Aurigæ	5 8.2	+45 53	0.2	−0.3
936	β Orionis	5 9.0	− 8 20	0.2	0.2
3147	α Lyræ	18 33.0	+38 41	0.4	0.0
213	α Ursæ Minoris	1 16.6	+88 42	2.2	−1.9
2500	β Ursæ Minoris	14 51.0	+74 38	2.1	0.6
3077	δ Ursæ Minoris	18 9.0	+86 37	4.3	1.1
1292	51 H. Cephei . .	6 46.7	+87 13	5.3	2.0
3426	λ Ursæ Minoris	19 37.9	+88 58	6.5	2.1
—	DM +89° 3 . .	2 38.8	+89 38	9.2	3.7
—	σ Octantis . . .	18 33.5	−89 16	5.8	1.1
—	DM +89° 37 .	18 24.0	+89 54	10.5	3.6

The conditions needed to photograph a faintly illuminated surface, as a nebula, are quite different. The four sources of error noted on page 187 will here have no effect on the result, except in obscuring details, unless the surface is very small. Using the same notation as before, $l = b \frac{f^2}{a^2 t T^2}$. Accordingly, equally faint surfaces may be photographed by similar lenses, whatever their size, except that a large lens has a slight disadvantage from the greater absorption due to the increased thickness. When, however, detail is to be shown, the advantage of the large scale of the images formed by a lens of long focus at once shows itself. An increase in the angular aperture is, however, a great advantage in photographing a faint surface.

When the different portions of a plate are subjected to a gradually increasing light, up to a certain point no perceptible effect will be produced. The darkening then becomes more and more intense, until a certain maximum effect is produced, and then with a very intense light, like that of the Sun, a reversing action takes place, by which the density becomes less and less. Representing by a curve the relation between the darkening and the total actinic energy received upon any portion of the plate, we find that these curves may differ in three essential particulars. First, the amount of light required to produce the first impression upon the plate. This may be regarded as a measure of the sensitiveness of the plate, or its value for photographing very faint objects. Secondly, one curve may be steeper than another; that is, the increase in darkening may be more marked with one plate than another, with a given increase of light. The greater this increase, the better is the plate adapted to show differences in the light of stars of nearly equal magnitude, or to show details in nebulæ or spectra. On the other hand, the range of such a plate is small, and it will be less adapted for making charts or other pictorial representations. Moderately bright stars will completely decompose the silver particles, and cannot be distinguished from much brighter objects. The brighter portions of a nebula or spectrum will also be burned out, and will fail to show great variations in light as well as plates less sensitive to small changes. Thirdly, the maximum darkness of different plates also varies; but this is a matter of less importance for our present purpose. A fuller consideration of this subject, with measures of the constants of various plates, will be given by Mr. W. H. Pickering in the Proceedings of the American Academy.

In view of the continual improvement in photographic processes, and the increase in sensitiveness that has been attained in the more recent forms of plates, it becomes an interesting question to consider what is likely to limit the results attained. We are already approaching this limit on moonlight nights. The fogging of the plates is so great when the moon is nearly full, that long exposures cannot be used with a telescope of so large an angular aperture as the telescope here employed. A further increase in the sensitiveness of the plates will render it impossible to work to the best advantage in the vicinity of a large city, on account of the illumination of the atmosphere by artificial light. It will then be necessary to take the photographs in more remote regions, or preferably at great elevations, where the reflecting atmosphere is diminished in amount. For these reasons a great increase in the time of exposure, or in the angular aperture of the telescope, is not to be desired. As shown above, with a given linear aperture the light of the sky or other luminous surface will diminish rapidly as the focus is increased. The light of a star will be diminished only so far as the diameter of its image is increased by the increase in focus. A great saving in time and expense might, however, be effected by more sensitive plates, since smaller lenses and shorter exposures could be employed.

In order that the stars shall leave trails, it is not necessary that the telescope shall be at rest. If its motion deviates in any way from that of the star, a trail will evidently be produced having a length proportional to the rate of deviation and to the length of exposure. If the speed of the driving-clock is greater or less than it should be, trails will evidently be formed having a length proportional to the hourly rate of the clock as compared with a sidereal clock, to the time of exposure, and to the cosine of the declination of the star. The light required to produce a trail of given intensity will bear the same relation to that required to produce the same trail when the telescope is at rest, as the hourly rate bears to an hour. Accordingly, if the clock gains or loses one minute an hour, trails will be formed by stars having one sixtieth part of the brightness of those which form similar trails when the telescope is at rest. This corresponds to a difference of about four magnitudes and a half. A rate of a second an hour should give the ratio of one to thirty-six hundred, or nearly nine magnitudes. The limit is soon reached, however, in consequence of the size of the images of the stars, and the impossibility of giving long enough exposures to enable them to traverse distances great enough to form an appreciable trail. The method of electrical control above described enables the rate of the driving-clock to be varied at will. The deviation from this cause can be rendered entirely insensible if desired.

Trails will also be formed if the axis of the instrument is not parallel to that of the earth. If we wish to make the images of the stars perfectly circular, instead of elongated into trails, this is a much more troublesome source of error. The simplest case to be considered is that in which we should give an exposure of twenty-four hours to the region in the vicinity of the north pole. The plate may here be regarded as revolving around a line passing through the optical centre of the object-glass and parallel to the polar axis of the instrument. The intersection of this line with the plate will form a centre around which the plate will appear to revolve. If a photograph should be taken of any fixed point in the sky, it would describe a circle around this point. If a star were situated exactly at the north pole, it would form such a fixed point, and would accordingly describe a circle with a radius equal to the distance between the pole of the earth and that of the instrument. Since the photographic surface sensibly coincides with a sphere whose radius is the focal distance of the lens, the deviation in seconds, s, of the axis may readily be determined from the radius r of this photographic circle. The relation between them will be expressed by the formula $r = \frac{sf}{206265}$, or $s = \frac{206265\ r}{f}$. No star is visible exactly at the pole, nor would it remain there, owing to precession. Adjacent stars will, however, describe nearly the same path, as will be shown below.

Since an exposure of twenty-four hours is impracticable, a portion only of the circle is obtained, the length of which is generally given with sufficient precision by the formula $l = \frac{m\ r}{229}$, in which l is the length of the trail and m the exposure in minutes. Combining this with the previous formula, we deduce $s = \frac{229 \times 206265\ l}{m\ f}$. The curvature of the trail is here neglected, but is readily allowed for if the exposure much exceeds an hour. If the exposure is much less than half an hour, it becomes difficult to detect the curvature, and to know on which side of the trail the centre of the circle lies; in other words, to decide whether the polar axis is too high or too low, to the east or the west. This difficulty may be remedied by stopping the clock and allowing each star to make a second trail for a minute or so by the diurnal motion. When the plate is developed, the trails formed when the clock was on will be parallel and of equal length; the others will be at right angles to a line drawn through the pole, and will have a length proportional to the polar distance. The second trail will also be fainter than the first, if the instrument is nearly in adjustment, except for stars very near the pole. The direction in which the first trail is described may be known from the fact that its following end is always attached to the preceding end of the second trail. Moreover, the direction in the northern hemisphere as seen from the object-glass is always opposite to that of the hands of a watch. Accordingly, if the observer holds the plate so that the first trail is horizontal with the second trail attached to its right-hand end, and the side of the plate to which the film is attached towards him, the centre of the trail will always be above. The simplest rule, however, is to make a second exposure after moving the axis, and to notice the effect on the trail. From this establish an empirical rule, and always hold the plates in the same position. When the trails are very short, it is sometimes better to detach them by covering the plate for a minute or so to prevent their interfering. Care must then be taken in making the exposures that the telescope is not disturbed.

We must next consider the trail described by stars in other portions of the heavens, supposing the only error to be that of the axis of the instrument. If it were possible to photograph a star at the south pole without disturbing the instrument, evidently it would also describe a circle like that described by a star at the north pole, since, if the photographs were taken simultaneously, both stars would always be exactly 180° apart. The path of other stars may be determined by conceiving of two concentric spheres revolving with equal velocity, the outer one around an axis parallel to that of the earth, the inner one parallel to the axis of the instrument. The motion of any point of the outer sphere compared with the inner one will give the required trails. Since the inclination of the axis is assumed to be small, the effect will be the same as if the inner sphere remained at rest and the outer one was moved so that its poles should describe circles around the axis of the inner sphere, but without rotating either sphere. This could be accomplished mechanically by loading the lower half of the inner sphere. If we consider a point upon the equator of the outer sphere, we see that its motion will be exactly north and south, any tendency of the northern pole to move it east or west being always compensated by an equal tendency of the southern pole to move it in the opposite direction. For other declinations, the east and west motion will be reduced by an amount that will always be proportional to the sine of the declination, while the north and south motion will be undiminished, and will always be equal to twice the deviation of the two axes. In the case of stars, it therefore follows that the trails will always be ellipses having their transverse axes north and south. The length of the semi-transverse axis will equal the distance between the pole of the instrument and the north pole, when represented on a sphere having a radius equal to the focal distance. The semi-conjugate axis at the pole will be the same as the semi-transverse axis; it will become zero at the equator, and will vary between these points as the sine of the declination. The trails will therefore vary from circles at the pole to lines running north and south on the equator. If errors are present, due to the rate of the clock as well as to the position of the polar axis, each will produce its effect independently. The form of trail may therefore be constructed geometrically from the principles described above. For testing the instrument, the pole is the best region, since errors in altitude and azimuth in the axis here enter with their full force. At the equator an error in altitude is insensible, as the star is there describing the end of a very elongated ellipse.

To adjust the axis, it may sometimes be found more convenient to replace the sensitive plate by a positive eye-piece with cross-lines. Directing the telescope upon a star near the meridian, if the axis points west of the true pole, the star will appear to be moving to the north, or will move below the horizontal cross-wire. It must then be brought back by one half the amount of the deviation, by moving the axis. The opposite effect will be produced if the axis is directed to the east.

The following discussion of the path described by stars at various declinations, when the axis is not properly adjusted, has been prepared by Professor Searle.

The effects resulting from an imperfect adjustment of the polar axis of any equatorially mounted telescope are partly indicated in treatises upon practical astronomy. Equations relating to the subject here to be considered may be found, for example, in Chauvenet's Spherical and Practical Astronomy, Vol. II., pages 375 and 378; but, for the present purpose, they may be somewhat simplified.

It will be convenient to give the names of instrumental poles and instrumental hour circles to the points of the celestial sphere towards which the ends of the polar axis of the telescope are directed, and the great circles passing through these points. One of the instrumental poles will be situated in the same celestial hemisphere with the star to be photographed. Let γ denote the distance of this pole from the nearer of the two celestial poles, and δ the declination, regarded as positive, of the star. Let p denote the corresponding distance of the star from the instrumental pole, always regarded as positive. Let t denote the angle between the planes of two hour circles, one, which may be called the fixed circle, passing through the instrumental pole, and the other through the star. Let v denote the corresponding angle between the fixed circle and the instrumental hour circle of the star, and consider t and v as equal to 0 when the celestial pole lies between the instrumental pole and the star, so that the star is crossing the fixed circle and $p = \frac{1}{2}\pi - \delta + \gamma$. Suppose t and v, at this time, to be increasing with the diurnal revolution of the star. Then

$$\cos p = \sin \delta \cos \gamma - \cos \delta \sin \gamma \cos t. \qquad (1.)$$
$$\sin v \sin p = \cos \delta \sin t. \qquad (2.)$$
$$\cos v \sin p = \sin \gamma \sin \delta + \cos \gamma \cos \delta \cos t. \qquad (3.)$$

It will be assumed, in order to simplify the inquiry, that the photographic plate is placed so that its plane is perpendicular to that of the fixed circle when $t = 0$. If γ is small, this will be nearly true in practice; in other cases, this will be the most convenient position for the plate if it is to depict the whole of the apparent path of the star with respect to the telescope. The easiest supposition respecting the inclination of the plate to the polar axis of the telescope will be that this inclination is equal to δ; that is, we may assume that the telescope is set to the declination of the star, without regard to error of adjustment. If the clock driving the telescope is correctly regulated, which is here assumed to be the case, the plane of the plate is constantly perpendicular to the plane of an instrumental hour circle, which makes the angle t with the fixed circle. The orthographic projection of this hour circle upon the plate is a straight line, which may be regarded as an axis of abscissas. The intersection of this line with the radius of the sphere perpendicular to it may be assumed as the origin; the point of the sphere to which the radius is directed is at the distance $\frac{1}{2}\pi - \delta$ from the instrumental pole, which will therefore be orthographically projected upon the plate at the distance $\cos\delta$ from the origin. The orthographic projection here employed has the advantage of simplicity; it will not, however, correctly represent the relative dimensions of different parts of the curve described upon the plate by the image of the star, unless γ is very small. But the present inquiry is confined to the general form of the curve, which may be derived as well from the orthographic projection as from one more strictly appropriate to the circumstances of the case. The gnomonic projection would probably best represent the actual curve.

The small circles formed by the intersections of the sphere with planes perpendicular to the axis of the instrument may be called instrumental parallels. Their orthographic projections upon the plate will be elliptical, but the parts of these projections near the origin will in ordinary cases differ little from straight lines. The projection of the instrumental parallel of the star will intersect the axis of abscissas at two points, the distances of which from the origin may be expressed by $\sin\left(\frac{1}{2}\pi - \delta - p\right)$ and $\sin\left(\frac{1}{2}\pi - \delta + p\right)$. The first of these quantities, which is equal to $\cos(\delta + p)$, will here be employed as one of the data indicating the form of the curve which is described upon the plate by the image of the star. The employment of the second quantity is never necessary, although, when $\frac{1}{2}\pi - \delta < \gamma$, it may sometimes seem more appropriate.

The least distance of the image of the star from the axis of abscissas will be expressed by $\sin(t - v) \sin p$. This will appear upon consideration of the case in which $\delta = 0$, and the inclination of the plate to the axis of the instrument will not affect the length of the projection, provided, as has been supposed, that the plate remains perpendicular to the instrumental hour circle at the angle t from the fixed circle.

The data thus assumed for the position of the image are not the customary rectangular co-ordinates, since the perpendicular denoted by $\sin{(t-v)} \sin p$ does not intersect the axis of abscissas at the same point with the corresponding projection of the instrumental parallel. Usually, however, these data will differ little from rectangular co-ordinates, and they will serve in all cases to represent the general course of the image of the star upon the photographic plate. The ordinary symbols for rectangular co-ordinates will therefore be adopted, so that $x = \cos{(\delta + p)}$, and $y = \sin{(t-v)} \sin p$.

From (1), $\cos p = \sin{(\delta - \gamma)} + \cos \delta \sin \gamma (1 - \cos t)$; this quantity progressively increases from $\sin{(\delta - \gamma)}$ when $t = 0$ to $\sin{(\delta + \gamma)}$ when $t = \pi$. The value of x must increase and diminish with that of $\cos p$, unless $\delta + p$ is negative or greater than π. But δ and p are always positive by supposition, and the greatest value of p is $\frac{1}{2}\pi - \delta + \gamma$, which can never exceed $\pi - \delta$; x therefore increases while t increases from 0 to π, returning through the same series of values as t increases from π to 2π, so that the value of x is the same for any two values of t equidistant from 0 or from π, no other value of x being identical with this.

By definition,
$$y = \sin{(t-v)} \sin p = \sin t \cos v \sin p - \cos t \sin v \sin p.$$

Hence, from (2) and (3),
$$y = \sin t \sin \gamma \sin \delta + \sin t \cos t \cos \gamma \cos \delta - \sin t \cos t \cos \delta;$$

and after reduction
$$y = \sin \gamma \sin \delta \sin t - \sin^2 \tfrac{1}{2} \gamma \cos \delta \sin 2t. \qquad (4.)$$

Differentiating this equation with respect to t, we have
$$\frac{dy}{dt} = \sin \gamma \sin \delta \cos t - 2 \sin^2 \tfrac{1}{2} \gamma \cos \delta \cos 2t.$$

From (1),
$$\frac{d\cos p}{dt} = \sin \gamma \cos \delta \sin t;$$

we have also
$$\frac{dx}{d\cos p} = \frac{\sin{(\delta + p)}}{\sin p} = \frac{\sin{(\delta + p)}}{\sin{(\delta + p - \delta)}} = \frac{1}{\cos \delta - \sin \delta \cot{(\delta + p)}}.$$

Accordingly,
$$\frac{dx}{dt} = \frac{\sin \gamma \cos \delta \sin t}{\cos \delta - \sin \delta \cot{(\delta + p)}},$$

and
$$\frac{dy}{dx} = \frac{[\cos \delta - \sin \delta \cot{(\delta + p)}] \, [\sin \gamma \sin \delta \cos t - 2 \sin^2 \tfrac{1}{2} \gamma \cos \delta \cos 2t]}{\sin \gamma \cos \delta \sin t};$$

this may be reduced to the form

$$\frac{d\,y}{d\,x} = \sin \delta \left[1 - \tan \delta \cot (\delta + p)\right] \left[(1 - \tan \tfrac{1}{2} \gamma \cot \delta \cos t) \cot t + \tan \tfrac{1}{2} \gamma \cot \delta \sin t\right].$$

In this expression $\sin \delta \left[1 - \tan \delta \cot (\delta + p)\right]$ cannot become negative, since δ and p are positive, while δ does not exceed $\frac{1}{2} \pi$, and when $\cot (\delta + p)$ is positive, $\tan \delta \cot (\delta + p)$ is a proper fraction. It reaches the value 1 only in the case of the passage of the star through the instrumental pole.

Equation (4) shows that y reaches the value 0 when

$$\sin \tfrac{1}{2} \gamma \sin t \left(\sin \delta \cos \tfrac{1}{2} \gamma - \cos \delta \sin \tfrac{1}{2} \gamma \cos t\right) = 0;$$

that is, when $t = 0$, when $t = \pi$, and when $\cos t = \tan \delta \cot \tfrac{1}{2} \gamma$; the last condition requires that δ shall not exceed $\frac{1}{2} \gamma$, and in that case will occur at some value of t from 0 to $\frac{1}{2} \pi$, and at the corresponding value between $\frac{3}{2} \pi$ and 2π. The points of the curve where these values occur are accordingly situated upon the axis, and the general statement of the variations of x, already given, shows that x has the same value at each point. Hence the two points are identical. It has been shown that, in the expression for $\frac{d\,y}{d\,x}$, $\sin \delta \left[1 - \tan \delta \cot (\delta + p)\right]$ is positive. The remaining factor, since $\tan \delta \cot \tfrac{1}{2} \gamma = \cos t$, is reduced to $2 \sin t \cos t$, which is positive for the smaller, and negative for the larger value of t. As $d x$ is also positive for the smaller, and negative for the larger value, $d y$ is positive in both cases. When $t = 0$, and $\delta < \frac{1}{2} \gamma$, $\frac{d\,y}{d\,x} = -\infty$; so also when $t = \pi$, $\frac{d\,y}{d\,x} = -\infty$.

Hence, when $\delta < \frac{1}{2} \gamma$, the curve consists of two closed branches with a common point, resembling a lemniscata; it will be shown below that, when $\delta = 0$, the equation of the curve represents a species of lemniscata.

When $\delta = \frac{1}{2} \gamma$, $\cos t = \tan \delta \cot \tfrac{1}{2} \gamma$ only when $t = 0$; the lower branch of the curve, accordingly, disappears. The term $(1 - \tan \tfrac{1}{4} \gamma \cot \delta \cos t) \cot t$ becomes $2 \sin^2 \tfrac{1}{2} t \cot t$, which may be written in the form $\sin \tfrac{1}{2} t (\cos \tfrac{1}{2} t - \sin \tfrac{1}{2} t \tan \tfrac{1}{2} t)$; when $t = 0$, this vanishes, and also the term $\tan \tfrac{1}{2} \gamma \cot \delta \sin t$; hence $\frac{d\,y}{d\,x} = 0$. As this value occurs at the minimum of x, the lower branch of the curve vanishes in a cusp. The value of p in this case is $\frac{1}{2} \pi + \frac{1}{2} \gamma$.

When $\delta > \frac{1}{2} \gamma$, $y = 0$ only when $t = 0$ or when $t = \pi$. When $t = 0$, $\frac{d\,y}{d\,x} = \infty$; when $t = \pi$, $\frac{d\,y}{d\,x} = -\infty$, as before. For values of t near 0, $\frac{d\,y}{d\,x}$ is relatively small as compared with the corresponding values when t is near π. The curve is therefore ovoid.

When $\delta = \frac{1}{2}\pi$, $\frac{dx}{dt} = 0$, and $\frac{dy}{dt} = \sin \gamma \cos t$; in this case $p = \gamma$, $x = -\sin \gamma$, and $y = \sin \gamma \sin t$. With the system of co-ordinates which has been employed, this represents the circle in which the celestial pole appears to move round the instrumental pole.

When $\delta < \frac{1}{2}\gamma$, an extreme value of y will occur for some value of t between 0 and $\frac{1}{2}\pi$. This is apparent from the values already found for which $y = 0$. The extreme values of this branch of the curve will be numerically greatest when $\delta = 0$, as is shown by (4), where the two terms of the value of y have contrary signs for values of t between 0 and $\frac{1}{2}\pi$. Their difference, accordingly, will be largest when the first term has its least and the second its greatest numerical value. This occurs when $\delta = 0$, and the value of y is then $-\sin^2 \frac{1}{2}\gamma \sin 2t$; at its extreme values, when $t = \frac{3}{4}\pi$ or $t = \frac{1}{4}\pi$, $y = \pm \sin^2 \frac{1}{2}\gamma$. The corresponding maximum and minimum in the other branch of the curve have in this instance the same value. The general condition for a maximum or minimum of y appears from the value of $\frac{dy}{dt}$ to be $\cos t \left(\cos t - \frac{1}{2}\cot \frac{1}{2}\gamma \tan \delta\right) = \frac{1}{2}$. This is satisfied by $t = \pm \frac{1}{4}\pi$, if $\delta = 0$, as has just been shown; also by $\cos t = 1$, if $\delta = \frac{1}{2}\gamma$; this result, also, has been considered above. In other cases, the values of $\cos t$ required for the maximum and minimum of y are found, by the solution of the quadratic equation just given, to be

$$\frac{1}{4}\left(\cot \frac{1}{2}\gamma \tan \delta \pm \sqrt{8 + \cot^2 \frac{1}{2}\gamma \tan^2 \delta}\right).$$

To find the greatest and least possible values of y, we have also, from (4),

$$\frac{dy}{d\delta} = \sin \gamma \cos \delta \sin t + 2 \sin^2 \frac{1}{2}\gamma \sin \delta \sin t \cos t,$$

which must vanish for the extreme values required, so that $\cos t = -\cot \frac{1}{2}\gamma \cot \delta$. Equating the two expressions thus found for $\cos t$, we have

$$\cot \frac{1}{2}\gamma \cot \delta = -\frac{1}{4}\left(\cot \frac{1}{2}\gamma \tan \delta \pm \sqrt{8 + \cot^2 \frac{1}{2}\gamma \tan^2 \delta}\right),$$

and, after reduction, $\tan^2 \delta = \frac{2}{\tan^2 \frac{1}{2}\gamma - 1}$. By supposition, γ is positive, and cannot exceed $\frac{1}{4}\pi$; hence $\tan^2 \frac{1}{2}\gamma$ never exceeds 1, and no extreme value of y can occur unless $\gamma = \delta = \frac{1}{2}\pi$. Accordingly, the numerical value of the maximum and minimum of y for a given value of δ increases from $\sin^2 \frac{1}{2}\gamma$ when $\delta = \frac{1}{2}\gamma$ to $\sin \gamma$ when $\delta = \frac{1}{2}\pi$, without reaching an algebraic maximum unless $\gamma = \frac{1}{2}\pi$, when $\sin \gamma$ denotes an extreme, as well as a final value.

No material modifications of the expressions already given are apparently required when the star passes between the instrumental and celestial poles. The equation of the curve in two special cases is given below.

When $\delta = 0$, the co-ordinates are rectangular;

Hence
$$x = \cos p = -\sin \gamma \cos t, \qquad \text{and} \qquad y = -\sin^2 \tfrac{1}{2} \gamma \sin 2\ t.$$

and
$$\cos^2 t = \frac{x^2}{\sin^2 \gamma}; \quad \sin^2 t = \frac{\sin^2 \gamma - x^2}{\sin^2 \gamma}; \quad \sin^2 2\ t = \frac{4\ x^2 (\sin^2 \gamma - x^2)}{\sin^4 \gamma};$$

$$y^2 = \frac{4 \sin^4 \tfrac{1}{2} \gamma}{\sin^4 \gamma} x^2 (\sin^2 \gamma - x^2) = 2 \sec^4 \tfrac{1}{2} \gamma\ x^2 (\sin^2 \gamma - x^2).$$

This is the equation of a curve in the general form of a lemniscata. If γ is sufficiently small, it may be reduced to $y^2 = 2\ x^2 (\gamma^2 - x^2)$, whence $x^4 - \gamma^2 x^2 = -\tfrac{1}{2} y^2$, and $x^2 = \tfrac{1}{2} (\gamma^2 \pm \sqrt{\gamma^4 - 2\ y^2})$. Accordingly, for any real value of x, y cannot exceed $\frac{1}{\sqrt{2}} \gamma^2$, and x cannot exceed γ; if γ is infinitesimal, y must be infinitesimal with respect to x, so that the lemniscata becomes a straight line.

In all cases, when γ is so small that we may substitute γ for $\sin \gamma$ and 1 for $\cos \gamma$, the difference between $\tfrac{1}{2} \pi$ and $\delta + p$, which never numerically exceeds γ, is a quantity of the same order; so also, accordingly, is x. Let $c = \tfrac{1}{2} \pi - \delta$;

$$\sin (c - p) = \cos \delta \cos p - \sin \delta \sin p = \cos (\delta + p) = x.$$

As x is small, we may also write $x = c - p, p = c - x$, and $\cos (c - p) = \cos x = 1$
Since
$$\cos p - \cos c = 2 \sin \tfrac{1}{2} (c - p) \sin \tfrac{1}{2} (c + p),$$
$$\cos p - \cos c = 2 \sin \tfrac{1}{2} x \sin (c - \tfrac{1}{2} x) = x (\sin c - \tfrac{1}{2} x \cos c).$$

Also, from (1), $\cos p - \cos c \doteq \cos p - \sin \delta = -\gamma \cos \delta \cos t$; hence $x = -\frac{\gamma \sin c \cos t}{\sin c - \tfrac{1}{2} x \cos c}$, where the term of the denominator containing x may be omitted, so that $x = -\gamma \cos t$. From (4), $y = \gamma \sin \delta \sin t$. Hence $\cos^2 t = \frac{x^2}{\gamma^2}$, $\sin^2 t = \frac{y^2}{\gamma^2 \sin^2 \delta}$, and $\frac{x^2}{\gamma^2} + \frac{y^2}{\gamma^2 \sin^2 \delta} = 1$. The curve is therefore an ellipse, becoming a circle of the radius γ when $\delta = \tfrac{1}{2} \pi$, and a straight line, as already shown, when $\delta = 0$, since in that case $y = 0$.

The general results of the inquiry, accordingly, are that, without restriction as to the amount of the error of adjustment, the curve described by the image of a star situated on the equator is a species of lemniscata; with an increase in the declination of the star, the lower branch of the curve becomes smaller and narrower, and disappears in a cusp when the declination is equal to half the error of adjustment. For greater declinations, the curve is ovoid, and at the pole it is circular. If the error of adjustment is sufficiently small, the curve is a circle at the pole, a straight line at the equator, and an ellipse in intermediate declinations, as has been stated on page 195.

TRAILS.

Various advantages accrue to the method of photographing the stars without moving the telescope. Each star as it passes through the field leaves a trail which appears on the plate as a fine line, forming part of a circle having the pole as a centre. The first advantage of such a line is, that it is distinguished with certainty from a defect in the plate. When the photograph is to be used as a measure of stellar magnitude, the trail shows that the plate has the same sensitiveness throughout. In ordinary plates this condition appears to be perfectly fulfilled. The trails appear as lines whose intensity is perfectly uniform within the limits of accuracy of which the comparison is capable. Small differences in light are much more perceptible in the trails than in the circular images formed when the telescope is driven by clockwork. The principal objections to photography as a means of determining the brightness of the stars are, first, that for slight differences in brightness the photographic images differ less than the real images. This objection is, however, counterbalanced by the possibility of repeating indefinitely doubtful measures, and of comparing a large number of similar trails under nearly the same conditions. Secondly, a variation in focus in different portions of the plate may affect the measures. A star out of focus may leave a broad trail and appear brighter than another which gives a narrow trail in consequence of its being more nearly in focus. The fact that the photographic intensity will vary greatly with the color can scarcely be called an objection. We wish to know the true relative intensities of the light of the stars, and not merely their relative brightness as judged by the eye. As long as the spectra of the objects compared are the same, that is, as long as the light of any given wave-length emitted by each bears the same proportion to the whole, all methods of measurement will give the same result. In other words, the relative intensity will appear to be the same, whether it is measured by the eye or by the sensitive plate. This is the more precise statement of the case which is commonly expressed by saying that the color is the same. When the spectra differ, and the colors are unlike, no single number will properly express the ratio of the two lights. The only true comparison is by a series of numbers which express the ratio of the light for each different wave-length. When, therefore, we say that a red and a blue star appear equally bright, we merely indicate that the entire radiation affects the eye equally. The visual result will not in general differ much from what would be attained if all the light had a wave-length .00006 cm., or 6000 ten-millionths of a millimetre. The photo-

graphic plate gives a more precise summing up of all the radiations, since no difference of color appears in the final picture, but the mean wave-length is not far from
4000 ten-millionths of a millimetre.   Accordingly, blue stars will appear comparatively much brighter in the photograph, and red stars brighter to the eye.   Their
relative light can be fully determined only by the comparison of the spectra, which
will be considered later.   Meanwhile the photograph furnishes an excellent test of
the color of a star, since on comparison with the visual brightness the stars which
are faint photographically may be assumed to be red, and the bright ones blue.   As
the difference amounts to several magnitudes, it furnishes a test much more sensitive
than that of the eye.   Again, this method is applicable to the faintest stars visible,
when the difference in color is quite imperceptible by any other means.

The first tests that were made of the photographic lens, and before it was
mounted equatorially, consisted in directing it to the pole, and photographing
the trails of the polar stars.   This is probably the best method of testing the
quickness of any given form of lens, plate, or developer.   It may be employed by
any photographer, as it is only necessary to turn the camera to the Pole-star and
leave it exposed for any convenient time, as half an hour.   On developing the
plate, the faintest stars shown measure the sensitiveness.   Varying either the lens,
plate, or developer gives us a means of studying the quality of each.

An excellent means of securing an automatic record of the cloudiness during
the night consists in exposing a plate in this way.   A long focus lens should be
used, but it need not be carefully constructed.   The slide should be opened in the
evening as soon as it is dark, and it must be closed before the morning twilight.
This may be done automatically by an alarm-clock.   On developing the plate, the
Pole-star will describe a circular arc having a length of 15° for each hour of exposure.   The time of passage of any clouds will be marked by interruptions of greater
or less length.   Such an instrument also forms a photographic watch-clock.   The
watchman must cover the lens at intervals for a minute or so, each of which will
be indicated upon the plate when it is developed.

When the positions of the stars are to be determined photographically, the
trails possess some especial advantages.   The edges are well defined, and the
errors introduced by the irregularities of the clockwork and the shaking of the telescope when in motion are avoided.   The declinations can be measured with greater
accuracy than the right ascensions.   For the latter it is best to make a number
of breaks at specified times, thus breaking up the lines into a number of dots whose
centres can be determined with accuracy.   Much care is necessary to avoid touch-

ing the telescope when covering or uncovering it, as the slightest flexure is sufficient to distort the ends of the trails. The measures are necessarily relative except
in the case of polar stars, whose absolute declinations may be determined if the
centre of the circles constituting the trails can be fixed with sufficient precision.

Unfortunately, the method of trails is not applicable to very faint stars unless
they are near the pole. A close polar star no brighter than the fourteenth magnitude gives a satisfactory trail, but equatorial stars fainter than the eighth magnitude
have not as yet been photographed in this way, on account of their rapid motion.

As stated above, short trails are produced whenever the telescope is not perfectly
adjusted. This is an objection to the appearance of the images upon a chart, and
prevents the faintest stars from forming images. On the other hand, the advantages
of distinguishing the images from defects in the plate, and the greater accuracy with
which the brightness can be measured, may render it advisable to employ this
method.

A wide field of work appears open in the application of photography to meridian
instruments, or to the almucantar. The sensitive plate should be substituted for
the reticule, and the position marked by a graver attached to the tail-piece of the
telescope. The times may be indicated by covering and uncovering the object-
glass automatically by the sidereal clock. The intervals should be determined by
trial, so as to give a series of nearly circular dots separated by as short intervals
as possible. Another method is to attach the plate to the armature of an electro-
magnet, the current being made and broken at regular intervals. Two series of
alternate lines of dots are thus formed. A large number of stars may be recorded
on a single plate. The principal advantage of this method would be its freedom
from personal equation. It is therefore especially adapted to longitude campaigns.
Moreover, a high degree of skill is not required by the observer. It would not be
necessary to employ a large telescope in this work. A 3-inch object-glass with a
focal length of 44 inches would give a sixth-magnitude star as well as an eighth-
magnitude star is shown in the Bache telescope.

A large number of photographs have been taken of the immediate vicinity of
the pole, both with and without clockwork. A special section of this memoir is
devoted to the discussion of a portion of them. (See page 218.)

The second research to be described is undertaken to determine the light of all
of the brighter stars. In order to include all portions of the sky, several exposures are made on each plate. The region to which each star belongs is indicated
by varying the time of exposure. Since the length of the trail is proportional

to this time, by giving two or more exposures to each region each star may be made to indicate the region in which it is situated by a series of characters like those of the Morse telegraphic alphabet. Each plate covers a region 10° square, and therefore extends over 40 minutes on the equator, and more in other declinations. Eight regions are taken on each plate in the first series. The settings are made at −20°, −10°, 0°, +10°, +20°, +30°, +40°, and +50°. The region covered therefore extends from −25° to +55°. One minute is devoted to each region. The first exposure lasts as many seconds as there are degrees of declination in the southern edge of the region. A break then occurs for ten seconds, and for the last five regions, that is, for those north of the equator, the second exposure lasts during the remainder of the minute. For the first three regions, a second break of ten seconds is made, extending from the fortieth to the fiftieth second. The regions covered in each case, the length of the exposures, and the appearance of the trails, are as follows: —

−25° to −15°	25 seconds,	5 seconds,	10 seconds, ⎯ · ·
−15° " − 5°	15 "	15 "	10 " ⎯ · ⎯ ·
− 5° " + 5°	5 "	25 "	10 " · ⎯ ·
+ 5° " +15°	5 "	45 "	· ⎯⎯
+15° " +25°	15 "	35 "	⎯ ⎯
+25° " +35°	25 "	25 "	⎯⎯
+35° " +45°	35 "	15 "	⎯⎯ ⎯
+45° " +55°	45 "	5 "	⎯⎯ ·

These plates are to be taken for every 40 minutes throughout the entire twenty-four hours of right ascension. Another series is taken for every alternate twenty minutes, and, to make these overlap still better, the declination is diminished for each of them by 5°. Accordingly, the centres of the regions in the second series coincide with the corners of the regions of the first series. Every star will therefore appear on at least two regions, and if near the corner of one, it will be near the centre of the other. Owing to the convergence of the meridians, some stars will appear on more than two plates. The total area to be covered is about 25,000 square degrees, and each plate will cover 800 degrees. As the series will contain 72 plates, the whole space will be covered on the average about two and a quarter times. As we go north, the lines become shorter, and therefore fainter stars will leave trails. At the northern limit the lines will have but little over one half their length on the equator, and stars half a magnitude fainter will appear. The most northern region will have one exposure of 45 seconds, which even at that declination will give a line long enough to be readily compared with the others.

Another series of photographs is made of the polar regions. These are taken at intervals of one hour in right ascension. Three regions are photographed on each plate. The first extends from +65° to +75°, and has an exposure of 30ˢ, an interval of 30ˢ, and an exposure of 120ˢ. The second extends from +75° to +85°, and has an exposure of 30ˢ, an interval of 30ˢ, and an exposure of 240ˢ. The third region extends from +75° to +85°, and is below the pole. The exposures are the same as in the first region, but as the stars are moving in the opposite direction, the short line now comes on the opposite side of the long one. The regions observed at lower culmination serve to correct the scale for portions of the sky differing by twelve hours in right ascension. They also serve. to determine the law regulating the atmospheric absorption. This is, however, much better determined by the next series of photographs. Three regions are photographed on each plate, all extending from +55° to +65°. The first is on the meridian above the pole, and is made by an exposure of 10ˢ, an interval of 10ˢ, and an exposure of 40ˢ. The second region has the same exposure, but is taken below the pole and at an hour angle of thirty minutes west, that is, it contains stars that have not yet culminated. The third region is also below the pole thirty minutes east. Two equal exposures of 30ˢ each are given, separated by an interval of 10ˢ. These plates are taken at intervals of twenty minutes in right ascension, and in general one is exposed each night. Two regions are photographed at lower culmination for one at upper culmination, because the stars form so much fainter images when low that comparatively few are obtained at each exposure.

The number of stars shown in these plates, especially in those relating to the polar regions, is very large. Even if it should prove impracticable to identify and measure them all, they will have a value as a permanent record of the condition of the sky at the present time. An illustration of this occurred in Plate 117, which was taken on November 9, 1885. It included the region in which the new star in Orion was discovered on December 13. No evidence of the new star is visible, although the adjacent stars DM. +19° 1106 and DM. +20° 1156 are so well shown as to be easily seen in a paper positive. The magnitudes of these stars in the Durchmusterung are 6.8 and 7.2 respectively. After allowing for the difference in color between these stars and the new star, it is evident that the latter must have been much fainter on November 9 than at the date of discovery. It is believed that no other positive evidence of this fact has been shown to exist.

CHARTS.

In the formation of charts of the stars by photography, we have a definite model to copy. It is not likely that any one will attempt to construct by eye observations charts of any considerable portion of the sky which will be more complete than those of Peters and Chacornac. If then charts equal to these can be obtained by photography, it may be regarded as an entirely satisfactory solution of the question. The area of these charts is 5° square, and their scale is 6 cm. to 1°, or three times the scale of the Durchmusterung. This scale corresponds to a focal length of 343.7 cm. or 135.3 inches. But it is impossible, without enlargement, to print the finest details visible on a good photograph, and, if printed, they could not be seen without a magnifying glass. The necessity of such a glass would greatly interfere with the general utility of star charts, especially when they are to be compared with the stars at night. Accordingly, the plan of enlarging the photographs does not seem objectionable, although some of the finer detail is lost. The scale of the photographs taken with the telescope described on page 184 is 2 cm. to 1°. If then they are enlarged three times, their scale will be the same as that of the charts named above. Lenses are made for ordinary photographic purposes which will include a field of view of 60°, or even 90°, without serious distortion. A photograph of the stars is, however, a far severer test. The distortion becomes perceptible even at a few degrees from the centre. With a single achromatic lens, the distortion is perceptible within a single degree; but with the compound achromatic, such as that of the telescope just mentioned, a much larger angle may be covered satisfactorily. The distortion at the sides of the plates, 5° from the centre, is not very large; at the corners of a plate 5° square, about 3°.5 from the centre, the errors are so small that they will not seriously affect the value of a map.

The advantages of this plan for constructing star charts are its economy and the rapidity with which the work can be performed. When several exposures are made on each plate, an error in one will ruin the whole. A single exposure of one hour is here proposed, which also diminishes the danger of interruption by clouds. The apparatus works automatically, and an observer is not needed who shall continually correct the motion of the clockwork by watching a star through an attached telescope. A great saving in fatigue is thus effected, and skilled labor is not required, since the work may easily be reduced to a routine.

The cost of continuing the work throughout the entire night would be small, since it would only be necessary for the observer to change the plate and readjust the instrument once an hour. If desired, the intervening time could be employed in other observations. The average length of a night, after allowing for twilight, is about ten hours. It would not be difficult to find a location where four nights in every week would be clear. This would give for the maximum capacity of a single photographic telescope nearly two thousand plates annually. The area covered by each plate is twenty-five degrees square. The total area of the sky is about forty thousand degrees square. Sixteen hundred plates would therefore be required to map the entire sky. Two stations must be employed to reach both northern and southern stars, and it therefore follows that it would be possible to prepare in this way a map of the whole sky in a single year. The final charts would not show the faintest stars that could be obtained by photography with larger instruments, but would give about as many stars in a given area as are contained in the charts of Peters and Chacornac. The charts should be carefully compared with the original negatives, to remove defects which might be mistaken for stars. To avoid the need of this comparison, the polar axis of the instrument may be moved slightly in azimuth. As shown on page 195, each star will then leave a short vertical trail. These can be distinguished with certainty from defects in the plate, and will give a more accurate indication of the brightness of the stars than can be derived from circular images.

STELLAR SPECTRA.

An investigation of the photographic spectra of the stars was conducted on an entirely different method from that employed by previous investigators, which has been described on page 182. A large prism was constructed, and placed in front of the object-glass, as was first suggested and tried by Father Secchi in his eye observations of stellar spectra.

The great advantages of this method are, first, that the loss of light is extremely small, and, secondly, that the stars over the entire field of the instrument will impress their spectra upon the plate. As a result, while previous observers have succeeded in photographing the spectrum of but one star at a time, and have not obtained satisfactory results from stars fainter than the second or third magnitude, we have often obtained more than a hundred spectra on a single plate, many of them relating to stars no brighter than the seventh or eighth magnitude.

The first experiments were made in May, 1885, placing a 30° prism in front of the object-glass of the lens described on page 182. No clockwork was used, the spectra being formed of the trails of the stars. In the spectrum of the Pole-star over a dozen lines could be counted. In the spectrum of a *Lyræ* the characteristic lines were shown very clearly. Exposures of two or three minutes were usually employed, although one minute gave an abundant width. In the spectrum of a *Aquilæ*, besides the lines seen in a *Lyræ*, some of the additional faint lines noticed by Dr. Draper were certainly seen.

In the autumn of 1885, two prisms were constructed, having clear apertures of 20 cm. and angles of about 5' and 15'. They could be placed over the object-glass of the photographic telescope without reducing the aperture. The second of these prisms was that actually employed in the experiments described below.

The prism was always placed with its edges horizontal when the telescope was in the meridian. The spectrum then extended north and south. If clockwork was attached, a line of light would be formed too narrow to show the lines of the spectrum satisfactorily. The usual method of removing this difficulty is the employment of a cylindrical lens to widen the spectrum; but if the clockwork is disconnected, the motion of the star will produce the same effect. Unless the star is very bright, the motion will, however, be so great that the spectrum will be too faint. It is only necessary to vary the rate of the clock in order to give any desired width to the spectrum. A width of about one millimetre is needed to show the fainter lines. This distance would be traversed by an equatorial star in about twelve seconds. The longest time that it is ordinarily convenient to expose a plate is about an hour. If then the clock is made to gain or lose twelve seconds an hour, it will have the rate best suited for the spectra of the faintest stars. A mean-time clock loses about ten seconds an hour. It is only necessary to substitute a mean-time clock for the sidereal clock to produce the required rate. It was found more convenient, however, to have an auxiliary clock whose rate could be altered at will by inserting stops of various lengths under the bob of the pendulum. One of these made it gain twelve seconds in about five minutes, the other produced the same gain in an hour. The velocity of the image upon the plate when the clock is detached could thus be reduced thirty or three hundred and sixty times. This corresponds to a difference of 3.7 and 6.1 magnitudes respectively. Since the spectrum of a star of the second magnitude could be taken without clockwork, stars of the sixth and eighth magnitudes respectively could be photographed equally well with the arrangement described above.

A number of photographs were taken of various portions of the sky, and to secure images of all the brighter stars the following system was also adopted.

Four exposures of five minutes each were made, setting the telescope at −10', 0°, +10', and +20'. The first of the two stops was used in each case, so that each spectrum had a width of about a millimetre. As the deviation of the prism is about 9', the centre of the region photographed at the first exposure had a declination of −19', and extended from −24° to −14'. The four exposures covered a region forty minutes of time in width, and extending from −24° to +16'. Thirty-six plates are required to complete this series, taken at intervals of forty minutes sidereal time, and beginning at $0^h 20^m$. Thirty-six more plates are taken at intervals of forty minutes, beginning at $0^h 0^m$, the declinations being diminished by five degrees. They cover the region from −29° to +11', and stars near the corners of the plates in the first series are near the centres of the second series of plates. The region from +11° to +56° is similarly covered by seventy-two plates, arranged as before, in two series. The settings in declination for the first of these are 30', 40', 50', and 60', and of the second series 25', 35', 45', and 55'. The length of each exposure is five minutes. Finally, the northern stars are included in thirty-six plates arranged in two series. The first of these contain three exposures, setting at declinations of 70', 80', and 90°, and at intervals in right ascension of one hour and twenty minutes. The right ascension of the centre of the first plate of this series is $0^h 40^m$. For the second series of eighteen plates, the declinations are diminished by 5'. In right ascension, they lie midway between those of the other series, the first being at $0^h 0^m$.

Photographs were also taken of the spectra of the fainter stars in certain regions. The auxiliary clock was set so that it should gain about ten seconds in an hour, and a single exposure of about an hour was made upon each plate. The work of photographing the entire sky by this process proved to be too large to be undertaken by the aid of the Bache Fund. Fortunately, Mrs. Henry Draper, as a memorial to her husband, has made provision for continuing this investigation at the Observatory of Harvard College. The results will therefore be described more fully elsewhere.

BRIGHT STARS IN THE PLEIADES.

As an example of some of the results to be derived from stellar photographs such as have been described above, the following examination has been made of several photographs of the Pleiades. The relative brightness of the principal components of this group has been determined from four photographs, Nos. 209, 248, 327, and 361. The first of these plates was taken on December 15, 1885. It was exposed for 5 minutes with the clock on, then the telescope was moved 1° in declination, and exposed for 1 minute with the clock on again. The clock was then stopped, and after an interval of 30 seconds a trail was formed for 30 seconds. Plate 248 was taken on January 6, 1886. An exposure of 10 minutes was given with the clock on, and this was followed by a trail of 1 minute. Plate 327 was taken on January 23, 1886, without clockwork. Eight exposures were given, having the lengths 10', 5', 2', 1', 1', 1', 0'.5, 0'.5, 0'.5, and 40' respectively. The objective was covered for a few seconds between each exposure. Unfortunately, in putting on the cap, a slight pressure on the telescope brought the images slightly out of line. Plate 361 was taken in a similar manner on February 9, 1886, but a much greater number of exposures of various lengths were made. From Plate 327, it appeared that some stars, the trails of which were visible when the exposure was sufficiently long, produced no effect with an exposure of a single second. The experiment was therefore tried with Plate 361, during a portion of the time, of making short breaks between exposures, lasting for several seconds. Measures could thus be made of the positions of much fainter stars.

A scale for measuring the relative intensities of these trails was constructed from No. 15. This photograph was taken, August 15, 1885, by pointing the telescope to the north pole, and giving a series of exposures of ten minutes each with different apertures. The clockwork was detached, so that the telescope was at rest, and the aperture was varied by inserting a series of circular diaphragms, having diameters of 2.07, 3.12, 4.92, 7.77, 12.33, and 19.55 cm. The last gave nearly the full aperture of the telescope, the others reduced it by quantities which respectively correspond to 4.86, 3.99, 3.00, 2.01, and 1.00 magnitudes. Each star accordingly left a trail consisting of six short lines connected together at the ends, and each representing a difference in light from that next it of one magnitude. The smallest aperture was however too large by about 0.1 cm., so that the interval between it and the preceding, expressed in magnitude, was only 0.87. The portion of the plate containing the trails of δ *Ursæ Minoris* and 24 *Ursæ Minoris* was cut out and attached to a piece of cardboard, so that it could be laid upon the image of the star to be measured. The brightest portion of the trail of δ *Ursæ Minoris* was assumed to represent the magnitude 4, and the faintest portion the magnitude 9. All of these results have been diminished by one magnitude to make the scale agree more nearly with that in common use. Each trail to be measured was compared directly with this scale, the fraction of a magnitude being estimated to tenths. After measuring in this way all the trails visible on each of the plates, the results were brought together, and every star found on either of the plates was looked for on all the others. Residuals were next taken from the mean of the measures, and exceeded three tenths of a magnitude in six cases only. One of these equalled five tenths, the others four tenths of a magnitude. Another estimate was made in each of these cases, with results which are given in the remarks following Table IV. The mean of this estimate and that originally made has been employed in the table. All of the measurements were made by Miss N. A. Farrar, and were entirely independent, even in the case of the repeated stars.

To determine the character of the spectra of the stars in the Pleiades, Plate 337 was taken on January 26, 1886. The exposure was 34 minutes, and the width of the spectra was about 0.06 cm. The spectra of nearly forty stars of this group are shown upon this plate, besides a large number of adjacent stars, since the region covered is ten degrees square. Nearly all the brighter stars in the Pleiades have a spectrum of the first type, in which the spectrum is covered by a series of well-marked lines at regular intervals, including the lines C, F, G, h, and H of the solar spectrum. The line K is wanting, or at least is too faint to be visible.

The results are given in Table IV., in which the stars are arranged in the order of their photographic brightness. Stars leaving trails on all four plates are given first, followed by those contained on three plates, two plates, and one plate, respectively. The first column gives the number of the star in the Catalogue of Wolf published in the Annales de l'Observatoire de Paris, Mémoires, Tom. XIV., Deuxième Partie. This is followed by the designation employed by Bessel for the star, the Durchmusterung number, the right ascension and declination for 1885, and the mean photographic magnitude. No correction is applied for the error in the diameter of the smallest diaphragm. The effect of diffraction would be to compensate for this error. The uncertainty in this quantity, and also in the limiting magnitude photographed, renders the magnitudes of the fainter stars somewhat doubtful. The next column gives the residuals in tenths of a magnitude, found by subtracting from the observed magnitudes on Plates 209, 248, 327, and 361 their mean values, negative residuals being represented by Italics. The next column gives in the same form the residuals found by subtracting the mean photographic magnitudes from the results of the authorities indicated by the letters at the head of the column. When the residual exceeds nine, it is indicated by a p if it is positive, and by an n if it is negative. The exact value is then given in the notes following the table. At the head of this column the letter a denotes the magnitudes found by Lindemann, Mémoires Acad. Imper. St. Pétersbourg, XXXII. No. 6, p. 22; b, observations of Pritchard, Monthly Notices Royal Astron. Soc., XLII., p. 227; *c*, observations in Uranometria Oxoniensis, p. 94; and *d*, observations in Harvard Observatory Annals, XIV., p. 398. To correct for the difference in the zero of the scale of these four catalogues, their magnitudes were first corrected by adding the quantities —0.07, —0.15, —0.12, and +0.10, respectively, before subtracting the photographic magnitudes. The sums of the positive and negative residuals for each catalogue were thus made nearly equal. The last column describes the spectrum. A denotes that the character of the spectrum is uncertain, owing to the presence of an adjacent star; F, that the spectrum is too faint to render its character certain, I, that the spectrum belongs to Class I. as described above; R, that the spectrum possesses some peculiarity more fully explained in the remarks following the table. These remarks also indicate which stars interfere when A is entered in the last column, and give additional observations when the first measures are discordant.

TABLE IV.

No.	Designation.	DM.	R. A. 1885.	Dec. 1885.	Magn.	Resbl.	a b c d	Sp.
			m. s.	*° ′*				
227	25 η Alcyone	+23° 541	40 38.9	+23 44.9	2.70	3 2 1 2	3 0 3 4	I
66	17 b Electra	+23° 507	38 2.9	23 44.8	3.82	2 3 0 2	2 2 0 1	I
115	20 c Maia	+23° 516	38 59.0	24 0.7	4.05	0 0 2 0	1 3 2 0	I
79	19 e Taygeta	+24° 547	38 21.7	24 6.3	4.05	2 0 0 0	2 4 4 5	I
349	27 f Atlas	+23° 557	42 19.4	23 42.1	4.45	1 1 1 1	3 . 6 6	I
146	23 d Merope	+23° 522	39 30.1	23 35.3	4.48	2 0 0 3	3 4 3 2	I
353	28 h Pleione	+23° 558	42 20.6	23 47.1	4.90	1 1 1 1	3 . 4 3	I
76	18 m	+24° 546	38 18.0	24 28.6	4.95	2 0 0 0	4 7 9 8	I
62	16 g Celæno	+23° 505	37 58.1	23 55.6	5.05	0 2 0 0	2 2 2 3	I
121	21 k Asterope	+24° 553	39 3.3	24 11.6	5.52	0 2 0 3	1 1 3 4	I
293	28	+22° 563	41 32.7	23 4.0	5.95	2 0 0 0	6 . 3 4	I
120	22 l	+24° 556	39 11.8	24 10.0	5.95	0 2 0 0	0 2 4 4	I
212	24 p	+23° 536	40 30.8	23 45.5	6.05	3 2 . .	1 4 1 .	I
182	12	+24° 562	40 8.2	24 9.7	6.32	2 0 3 2	1 2 3 4	I
406	34	+23° 563	42 54.3	23 21.6	6.42	2 1 3 1	6 . 3 2	I
423	33	+23° 569	43 8.3	23 29.9	6.52	0 0 3 2	0 . 2 2	I
105	4	+23° 512	38 47.3	23 58.4	6.52	0 0 2 3	3 9 8 .	I
300	29	+23° 553	41 39.1	23 59.5	6.60	1 3 2 2	1 1 1 1	I
370	32	+23° 561	42 30.1	24 1.7	6.62	1 1 4 1	2 . 4 2	I
338	26 s	+23° 556	42 7.1	23 30.3	6.70	2 1 3 2	7 4 3 2	R
226	21	+23° 540	40 38.9	23 56.0	6.80	2 2 0 0	5 5 4 2	A
151	10	+23° 523	39 36.9	23 53.7	6.82	2 0 3 2	1 2 2 2	I

TABLE IV. — *Continued.*

No.	Designation.	DM.	R. A. 1885.	Dec. 1885.	Magn.	Resid.	a b c d	Sp.
			m. s.	° '				
473	40	+23° 570	44 2.3	+23 36.8	7.02	0 2 3 0	6 . 0 3	I
141	8	+23° 519	39 23.6	23 50.1	7.40	1 1 1 1	4 1 2 .	A
214	20	+24° 566	40 32.5	24 13.9	7.40	1 1 1 1	3 . 0 0	I
213	19	+23° 537	40 31.8	23 26.8	7.42	1 0 1 1	8 8 8 6	A
217	22	+23° 538	40 34.9	23 33.5	7.48	0 1 0 0	6 9 8 6	A
447	39	+24° 578	43 36.5	24 8.7	7.48	0 1 0 0	3 . 3 3	R
418	37	+23° 567	43 5.3	23 59.9	7.48	0 1 0 0	2 . 3 3	I
365	31	+23° 560	42 25.5	24 2.6	7.52	0 2 0 0	2 9 8 .	I
235	25	+23° 542	40 48.1	23 15.2	—	. . . .	. . . .	A
358	30	+23° 559	42 22.5	23 32.1	7.52	0 3 2 0	2 1 2 .	A
208	17	+23° 535	40 28.8	23 22.1	7.55	1 2 1 2	u n 8 .	I
120	7	+23° 517	39 2.5	23 40.7	7.58	2 1 1 1	1 8 2 .	I
209	18	+23° 536	40 29.1	23 46.9	7.65	1 2 2 1	5 2 2 .	A
192	13	+23° 528	40 14.5	23 38.3	7.78	0 2 0 3	2 2 3 .	I
280	27	+23° 549	41 22.7	23 57.8	8.00	0 0 0 0	1 . . .	R
219	21	+24° 567	40 35.0	24 18.0	7.63	0 2 1 .	6 3 2 .	A
376	33	+23° 562	42 35.1	23 53.7	7.70	1 2 . 1	0 6 n .	I
91	1	+23° 510	38 36.4	23 40.4	7.70	1 2 . 1	2 6 4 .	I
101	2	+23° 550	38 43.7	24 6.1	7.90	. 1 2 1	2 . . .	F
198	14	+23° 530	40 21.3	23 25.4	7.93	1 . 1 1	9 . . .	F
11	—	+23° 495	36 53.7	24 0.7	7.93	1 1 . 1	. . . .	I
161	—	+23° 524	39 45.9	23 15.8	8.00	0 0 . 0	. . . .	I
407	35	+23° 564	42 55.2	23 53.6	8.07	1 . 1 1	p p p .	F
37	—	+23° 500	37 23.3	24 2.4	8.07	1 1 . 1	. . . .	F
109	6	+23° 513	38 51.1	23 55.6	7.40	. 1 1 .	p p p .	F
143	9	+23° 520	39 25.7	23 49.8	7.50	0 0 . .	2 2 1 .	A
225	23	+23° 539	40 38.1	23 19.3	7.75	. 1 0 .	4 5 3 .	I
202	15	+23° 531	40 26.6	23 46.3	7.75	3 2 . .	5 6 2 .	F
103	3	+23° 511	38 45.7	23 43.3	8.00	. 0 .	p . . .	F
204	16	+23° 533	40 27.5	23 27.6	8.00	. . . 0	. . . .	F
403	—	+24° 573	42 51.9	22 59.5	8.00	. 0 . .	. . . .	I
415	36	+23° 565	43 3.9	23 52.0	8.20	. 0 . .	8 4 8 .	F
4	—	—	36 40.9	24 17.5	9.00	. 0 . .	. . . .	F
45	—	—	37 33.3	23 8.9	9.00	. . . 0	. . . .	F

REMARKS.

146. Plate 361. A second measure gave 4.5 instead of the original measure 4.0. The mean value 4.2 has been adopted. The original residuals were 3 1 1 4.

212. Plates 327 and 361. The trails could not be measured on account of the proximity of 227.

182. Plate 327. A second measure gave 6.0 as before.

300. Plate 327. A second measure gave 6.5 instead of 7.0. The mean value 6.8 has been adopted. The original residuals were 1 3 4 2.

370. Plate 327. A second measure gave 7.0 as before.

338. Spectrum Type I, but the K line is as intense as the H line.

226. Plate 243. A second measure gave 7.0 instead of 6.2. The mean value 6.6 has been adopted. The original residuals were 3 5 1 1. Spectrum uncertain; 227 interferes.

141. Spectrum probably Type I, but not separated from 143.

213. Spectrum uncertain; 208 interferes.

217. Spectrum uncertain; 227 interferes.

447. Spectrum Type J, but the K line is visible, and about 0.2 as intense as the H line.

235. 227 interferes.

358. Plate 209. A second measure gave 7.0, but a more careful estimate gave 7.5 instead of the original measure 8.0. The mean value 7.5 has been adopted. The original residuals were 4 2 3 1. Spectrum uncertain; 349 interferes.

208. Residuals in a, 11; in b, 10.

120. Spectrum Type I. Presence of K line doubtful on account of interference of 115.

209. Spectrum uncertain; 212 interferes.

192. Spectrum Type I, but too faint to decide whether the K line is present.

280. Spectrum faint; lines narrow, if present.

219. Spectrum uncertain; 214 interferes.

376. Residual in c, 10.

101. Spectrum not seen; 105 and 115 interfere.

198. Spectrum not seen; 213 and 217 interfere.

11. Spectrum faint.

407. Residuals in a, 12; in b, 14; in c, 15.

109. Spectrum not seen; 105 interferes. Residuals in a, 16; in b, 17; in c, 19.

143. Spectrum probably Type I, but not separated from 141.

202. Spectrum not seen; 212 interferes.

103. Spectrum not seen; 105 interferes. Residual in a, 12.

204. Spectrum not seen; 213 and 217 interfere.

The spectra of Nos. 169 and 245 were seen, but were too faint to indicate their type. No trails were given by the stars 26, 11, and 5 *Pleiadum*, which have the numbers 245, 169, and 107 in Wolf's Catalogue.

An important inference may be drawn from the comparison of the spectra of the stars of this group. It is extremely improbable that chance alone has brought together so many bright stars in the same portion of the heavens. Most of them probably have a common origin, and are much nearer to each other than to the Solar System. A few, doubtless, have only an apparent connection with the group, their real distance being much greater or less than that of the others. Ordinary means fail to distinguish the individuals of these two classes. The similarity in the chemical and physical conditions indicated by the apparent identity of most of the spectra, is a strong confirmation of their common origin. The variation in the spectra of such stars as Nos. 338 and 447 seems to indicate that these stars happen to lie in the same direction from us as the others, but are not really connected with them. In a study of the parallax of the Pleiades, it seems very desirable that these stars also should be observed.

To determine the probable error of a single determination of brightness, those stars only should be included which are measured on all four plates. The original uncorrected residuals are also used even in those cases where a second measure showed that the first estimate was erroneous. The 140 residuals of the 35 stars included in this list give an average deviation of ±0.119 magnitudes. Using the corrected residuals, the average deviation would be reduced to ±0.106. The probable error of a single estimate will be 0.119 × 0.976, or ±0.12, and the error of a single star ±0.06. Of the residuals, 47 have the value 0; and of the positive residuals, 23, 15, 8, and 3 have the values +0.1, +0.2, +0.3, and +0.4, respectively. Of the negative residuals, 23, 13, 5, 2, and 1 have the values —0.1, —0.2, —0.3, —0.4, and —0.5. An examination of the entire number of 163 estimates showed that the frequency with which the various tenths of a magnitude were employed varied greatly; the figure 5 occurred 61 times; 0, 47 times; 8, 33 times; 3, 15 times; 2, 7 times; 4, 5 times; 7, 4 times; 6, once; 1 and 9, not at all. A test of the relative sensitiveness of the plates is afforded from the average value of the residuals relating to the images contained on each. The means for Plates 209, 248, 327, and 361, are +0.046, —0.040, +0.031, and —0.009. Accordingly, the stars on Plate 209 appear on the average about one twentieth of a magnitude fainter than on the average of the four plates. These corrections are so small that it is hardly necessary to apply them.

Measures were made of the positions of the images on Plate 327, to determine the degree of precision to be expected from the application of photography to transit instruments. The shortest exposures do not seem to have been much less than a second, owing to the time required to cover and uncover the glass. The images formed by the shorter exposures were minute circular dots, which for the fainter stars did not exceed 0.008 cm. in diameter. They were not perfectly symmetrical, probably owing to the diffraction when the lens was nearly covered. All the stars were, however, subjected to the same conditions. About a dozen of the brighter stars formed images even with the shortest exposures. The intervals between these images were measured by attaching the photograph to a micrometer-screw having a pitch of one twenty-fourth of an inch, by which it could be moved across the field of a microscope. The magnifying power was only ten diameters, and could doubtless have been increased with advantage. Eight settings were made on each star, seven on the centre of the dots formed by the exposures of two seconds, of one second, and of half a second, and one on the end of the last trail. Nine stars were measured in this way, and the first of them was taken a second time to see if the instrument had moved. The intervals were next found by taking the first differences of these readings. The means of the ten values of these seven intervals, expressed in revolutions of the micrometer-screw, were 0.380, 0.230, 0.139, 0.353, 0.145, 0.142, and 0.407. Each revolution is equivalent to 13".87. The residuals found by subtracting these mean values from the observed intervals are given in Table V., expressed in seconds of time. As in Table IV., the first column serves to designate the star by its number in the Catalogue of Wolf.

TABLE V.

No.	Designation.	Residuals.						
		I.	II.	III.	IV.	V.	VI.	VII.
849	Atlas	+0.06	+0.07	−0.03	−0.03	−0.01	0.00	+0.01
353	Pleione	0.00	+0.02	−0.03	−0.04	+0.10	+0.01	−0.06
146	Merope	+0.10	−0.04	−0.04	−0.01	+0.07	−0.03	+0.10
66	Electra	0.00	−0.06	+0.03	+0.08	−0.03	+0.04	0.00
62	Celæno	−0.13	−0.01	+0.13	+0.01	+0.04	−0.11	+0.03
115	Maia	+0.06	+0.03	−0.08	+0.01	−0.04	+0.04	+0.63
79	Taygeta	−0.21	+0.07	−0.01	−0.04	−0.01	+0.07	−0.11
121	Asterope	+0.11	−0.17	+0.10	−0.01	+0.01	−0.08	−0.07
227	Alcyone	+0.14	+0.03	−0.06	+0.03	−0.10	+0.08	+0.11
349	Atlas	−0.06	+0.01	0.00	+0.03	−0.03	+0.01	−0.04

The average values of these residuals, taken without regard to sign, are 0.087, 0.051, 0.051, 0.029, 0.044, 0.047, and 0.059, in the seven columns respectively. The first of these values is decidedly greater, showing that it was difficult to set with accuracy on the centre of so long a line as that produced when the exposure lasted for two seconds. The last value is also somewhat greater, owing to the difficulty of determining the end of a trail. The difference is, however, much less than was anticipated. The mean of all the residuals is 0.053, or, if we use only the intervals between short exposures and reject the first and last values, 0.044. Since all of these values are found from the difference of two settings, we must divide by the square root of two to obtain the average deviation of a single setting. We thus obtain the values 0.037 and 0.031. The probable error of a single setting is found by multiplying these values by 0.85, which gives ±0.031 and ±0.026. No setting has been rejected for discordance, and no change made in the original record except that the seventh setting on Maia was recorded 54.685 and was assumed to mean 54.635. This setting 54.685 would render the preceding and following values so large that they would have to be rejected for discordance. One would be increased and the other diminished by 0″.70. As shown in the Table, the assumed value gives the residuals +0″.04 and +0″.03. To show how far the deviations were due to errors of setting, ten successive settings were made on a single image of Celæno, and gave an average deviation of ±0″.015 and a probable error of ±0″.013. These various values could doubtless be reduced by the use of a higher magnifying power, that employed being much too low. The probable error of a transit over a single wire by the usual method may be taken as ± 0".06. The above measures show that the probable error of a single setting on the images of different stars is not far from one half of this, ± 0".03. The probable error of successive settings on the same image is only ± 0".015. The observations on different stars seem fairly to represent the accuracy to be expected in determining the position of stars by photography from transit observations. Besides errors of setting, there are included in the quantity above mentioned, 0".03, those due to lack of symmetry of the images, variation in the brightness of the stars, unequal expansion or irregularities in the film, so far as these could affect measures over small distances, and the various errors in the measuring instrument. Many of these quantities can doubtless be greatly diminished, but the results already obtained seem to prove the possibility of measuring the position of stars photographically from their transits, with an accuracy at least equal to that obtained in the usual manner.

CLOSE POLAR STARS.

A large number of photographs were taken of the vicinity of the north pole. Some of these were made when the telescope was at rest, the stars leaving trails; others, principally intended for testing the position of the polar axis, were made with the telescope driven by the clock. The number of stars on each plate may be inferred from the fact that over one hundred stars within one degree of the pole leave trails when the telescope is at rest. As each photograph extends five degrees from the pole, the complete reduction would be very laborious. The photographic brightness of the stars within one degree of the pole has been measured on three of the plates, Nos. 7, 21, and 231. The method of measurement was the same as that employed for the Pleiades and described on page 211. The attention of the observer, Miss Farrar, was called to the preponderance, in the estimates of the light of the Pleiades, of the zero tenths and five tenths of a magnitude, and this difficulty is corrected in the present measures. Plate 7 was taken on August 8, 1885, and had an exposure of 70 minutes; Plate 21 was taken on September 3, 1885, with an exposure of 165 minutes, and Plate 231 was taken on December 24, 1885, with an exposure of 50 minutes. The accordance of the measures is very satisfactory. The total number of stars is 117, and of the separate measures 330. The average value of the residuals found by subtracting the mean value of the brightness of each star from the separate values, expressed in magnitude, is 0.088. Accordingly, single measures of photographs taken on different nights will differ on the average by less than a tenth of a magnitude. It is doubtful if this degree of accuracy can be attained by any other photometric process hitherto employed. In no case does a residual exceed three tenths of a magnitude, and in one case only do the measures of the same star differ from one another by as much as six tenths of a magnitude. The greater portion of these stars are too faint to be contained in existing catalogues. Their positions have not yet been determined from the photographs. A full publication of their magnitudes does not therefore seem desirable in the present Memoir. As examples, the 38 stars enumerated in the Durchmusterung as belonging to this region have been selected from the entire list of 117 stars, and their measurements are given in Table VI. The successive columns give the Durchmusterung numbers, the photographic brightness of the trail, and the residuals from the mean expressed in tenths of a magnitude, negative residuals being indicated by Italics. In order to make the magnitudes correspond with those of equatorial stars giving trails of equal intensity, each of the original readings has been diminished by three magnitudes. By means of Table II. and the formulas given on page 190, a correction has been applied for the declination of

TABLE VI.

No.	Trail.	Resid.	Magn.	DM.	Res.	No.	Trail.	Resid.	Magn.	DM.	Res.
1	4.00	1 1 0	10.1	9.5	5	20	6.30	1 1 2	11.0	9.5	4
2	5.60	0 1 0	10.2	9.2	2	21	4.00	0 0 0	8.8	8.8	4
3	3.53	0 3 2	9.0	8.8	2	22	5.03	1 2 0	9.3	8.7	3
4	5.20	1 1 2	10.4	9.4	0	23	6.00	0 0 0	10.9	9.4	5
5	6.40	1 0 1	11.3	9.5	—	24	6.50	0 0 0	11.3	9.5	7
	6.40	1 0 1	11.3			25	3.93	0 0 1	9.1	9.0	5
6	5.63	0 3 2	10.2	9.4	2	26	4.87	0 1 0	10.9	9.5	3
7	4.77	3 2 0	9.4	9.3	8	27	6.27	1 1 1	10.6	9.5	0
8	6.70	1 1 0	11.5	9.5	9	28	5.17	2 1 2	9.7	8.7	7
9	4.93	1 1 1	10.0	9.1	2	29	4.27	3 0 2	9.9	9.4	5
10	5.80	1 1 0	10.3	9.5	3	30	5.43	2 0 1	10.0	9.3	2
11	6.03	0 1 0	10.4	9.5	2	31	5.57	2 1 0	11.2	9.5	6
12	4.70	1 1 0	10.0	9.1	2	32	5.70	2 0 2	10.4	9.5	2
13	2.77	3 1 1	7.1	7.0	—	33	4.93	3 0 2	10.1	9.5	5
14	6.27	0 1 0	10.6	9.5	0	34	5.60	2 0 2	10.4	9.3	2
15	5.47	1 1 1	0.8	9.2	2	35	4.67	1 0 1	10.8	9.3	6
16	6.13	0 1 0	11.1	9.5	5	36	4.27	3 0 2	9.8	9.5	8
17	4.23	1 2 0	9.1	9.0	5	37	3.93	1 1 1	11.0	9.3	8
18	4.83	3 1 3	9.6	8.9	2	38	4.77	0 1 0	9.4	9.0	2
19	5.77	1 1 1	10.7	9.5	1						

the star. The result, which is given in the fourth column, shows the actual photographic brightness of these stars. The fifth column gives the Durchmusterung magnitude, and the final column gives the residual found by subtracting the photographic from the DM. magnitudes, after reducing the scale of the latter to that of the former. Positive residuals indicate stars of a bluish color, and negative residuals those of a reddish color, provided no errors are present.

No. 5, or DM. +89° 5, is shown to consist of two equal stars, whose combined light would equal that of a star of the magnitude 10.6, and give a residual zero in the last column. Nos. 14 and 15 are so nearly in the same declination that their trails coalesce, although separate at the ends. Their combined light was originally measured, and the measures given were made subsequently. No. 27 was originally omitted, since by precession it is now outside of the limit of one degree north polar distance.

The greater portion of the stars in Table VI. differ but little in brightness. Accordingly, measurements were made of the polar stars proposed as standards of

TABLE VII.

Desig.	α 1880.	δ 1880.	Trail.	Resid.	Photog. Magn.	Photom. Magn.	Res.
	h. m.	° ′					
88° 8	1 14	88 40	—	. . .	—	2.2	—
86° 269	18 11	86 37	1.17	2 1 0	4.4	4.3	1
87° 51	6 44	87 14	2.40	2 1 1	5.7	5.3	4
88° 112	19 44	88 57	3.43	1 1 1	7.8	6.5	11
88° 4	0 51	88 23	3.08	1 1 1	6.8	7.0	2
88° 9	2 3	88 36	3.97	0 0 1	8.1	8.6	5
89° 3	2 28	89 36	3.37	1 1 1	8.9	9.2	3
89° 35	17 50	89 48	4.33	1 0 0	10.4	9.8	6
88° 37	19 28	89 54	3.67	1 1 1	10.8	10.5	3
89° 1	0 19	89 45	3.87	1 1 1	10.0	10.5	5
89° 26	13 23	89 49	5.13	2 0 1	11.1	10.6	5
a	19 30	89 54	6.00	0 0 0	12.9	12.2	7
b	19 30	89 55	5.93	1 1 1	13.0	12.4	6
h	22 0	89 50	6.23	2 1 2	13.2	12.8	4
k	23 10	89 53	6.13	1 1 1	12.7	13.2	5
d	14 0	89 57	—	. . .	—	13.3	.
l	0 0	89 54	—	. . .	—	14.0	.
c	18 30	89 58	6.60	2 0 1	14.2	14.0	2
e	9 10	89 58	—	. . .	—	14.8	.
f	3 4	89 58	—	. . .	—	14.8	.
g	0 10	89 57	—	. . .	—	15.7	.

brightness in the Proceedings of the American Association, XXXIII. 8. Their designations, right ascensions and declinations for 1880, and photometric magnitudes, are estimated from that publication, and are given in the first, second, third, and seventh columns of Table VII. The mean magnitudes of the photographic trails are given in the fourth column, and the residuals in the fifth column. The sixth column gives the magnitude corrected for declination, and the last column gives the photometric minus the photographic magnitudes. Positive residuals indicate blue, and negative, red stars. The first four stars are a Ursæ Minoris, δ Ursæ Minoris, 51 Cephei, and λ Ursæ Minoris. The trail of the first of these is too intense to be measured.

A second reading of the trail of b, Plate 231, gave 9.3 instead of 8.8. The mean value 9.0 has been adopted. The original residuals were 2 1 4. The red color of λ Ursæ Minoris is indicated by the large negative residual, the value of which is 13.

The readiest method of publishing the results obtained by any photographic process is to reproduce them so far as possible in paper prints obtained exclusively by photographic means from the original negatives. In such prints, however, as is well known, many of the details of the negatives cannot be traced, and an unfavorable impression of the value of the work may thus be occasioned. On the other hand, if every detail which may be detected in a negative is described, or reproduced by engraving, the original may be supposed to be far superior to other photographs of nearly the same actual value. In the present case, no attempt at an exhaustive study of the negatives has yet been made. What is here described may be seen upon them with little difficulty.

Plate II. represents the central portion of photograph No. 6, taken on August 6, 1885, with an exposure of 72 minutes, and enlarged five times. It therefore represents the portion of the sky within about one degree of the pole. The scale is five times that of the Durchmusterung, one degree equalling 10 cm. The stars contained in the Durchmusterung zone +89° are here designated by their numbers in that Catalogue. The designations of the other stars proposed as standards of magnitude for faint polar stars, and given in Table VII., are also inserted. The trails left by the stars are untouched, although, owing to the defects of the photolithographic process, they are very irregular. In the original they form perfectly smooth lines. The Pole-star appears as a broad band in the left-hand lower corner. The Voigtländer No. 4 lens, described on page 183, gave nearly as many stars as the larger lens afterwards employed. DM. +89° 37, magnitude 10.5, was always well shown, and under favorable circumstances star b, magnitude 12.4, was distinctly

seen. With the eight-inch lens DM. +89° 37 and a were distinctly separated in
the original negative. This is a satisfactory proof of the good definition of a lens
of so short a focus.

MISCELLANEOUS.

The different photometric processes which have been used in determining the
light of the stars give results which differ systematically when large variations in
light are to be measured. In the Harvard Observatory Annals, Vol. XIV. p. 504,
it is shown that the systematic variations of the best catalogues exceed one fifth
of the entire intervals measured. In other words, if a great number of pairs of stars
are selected, whose average difference in brightness is five magnitudes according to
one catalogue, the difference will be only four magnitudes in another. This is also
true when the logarithms of the light are used. It is not due to a difference in the
assumed value of the unit of the scale of magnitudes, but to an actual difference
in the measurement of the amount of light. In the case assumed above, if the light
of the brighter star is taken as 100, that of the fainter star according to one catalogue will be 1; according to the other, it will be 2.5. Evidently, large errors affect
one or more of the catalogues compared, which cannot be eliminated by increasing
the number of observations. The best way of determining which is probably correct
is to repeat the measures by a variety of entirely different methods. Photography affords an excellent means of doing this, since the errors, if any, will be of a
very different kind. Several methods may be employed, each giving an independent
test. The construction of a standard scale, as described on page 211, gives a direct
measure of the ratio of the light of two stars of the same color, if we can assume
that the brightness of the image is proportional to the area of the object-glass. The
principal objection to this method arises from diffraction, which enlarges the images
when the aperture is very small. A defect in the portion of the plate on which the
standard is photographed might affect all of the measures. This should be tested
by measuring all the stars on a plate which has received several exposures with
various apertures. The results also give a good means of studying the effect of
aberration at different distances from the centre of the plate. Instead of varying
the aperture, we may vary the time of exposure. This may be accomplished by
varying the rate of motion of the image. For stars in the vicinity of the pole, the
velocity is proportional to the polar distance. If, as on page 187, we assume that
the brightness of a star capable of producing a given impression will vary as the

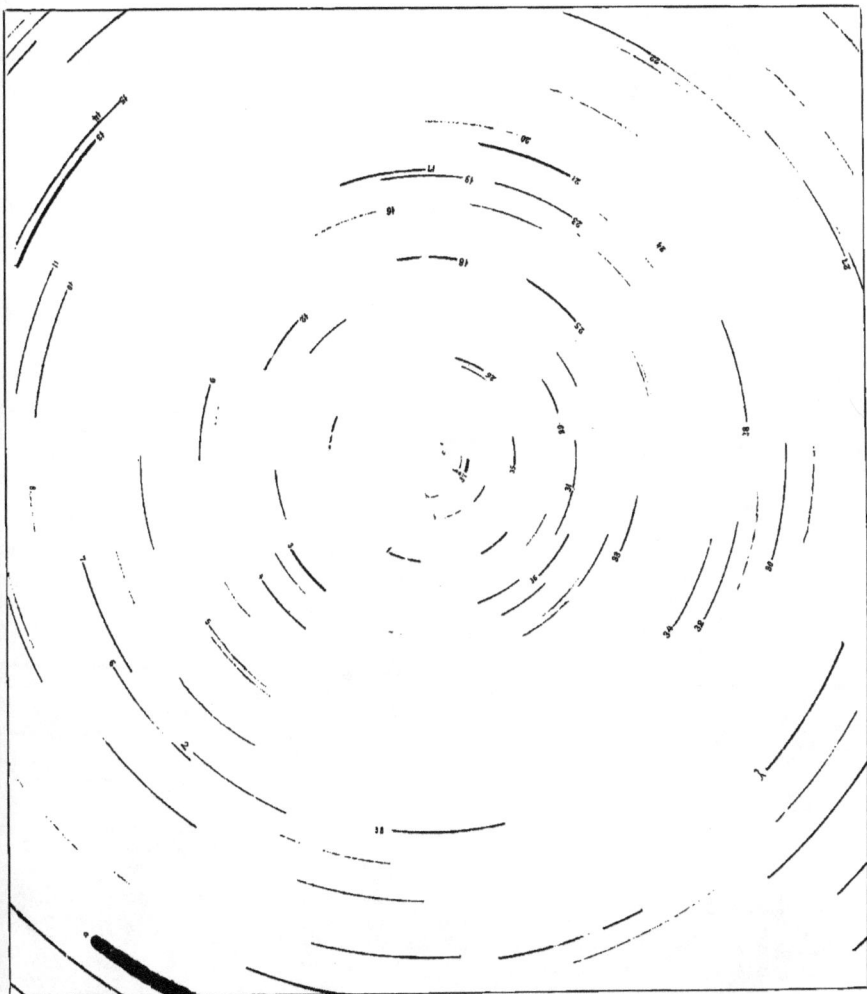

Plate 2 $r = 10$ cm.

— TRAILS OF CIRCUMPOLAR STARS. —

time of exposure, we may hence obtain a method of determining relative intensities of light. If two stars give equally distinct trails, their intensities must be inversely as their polar distances. The last column in Table VII. shows that this condition is nearly fulfilled. The difference in the photographic magnitude of 86° 269 and e is 9.8 magnitudes, and the photometric difference is 9.7 magnitudes. The image of the first star in this case moves nearly seventy times as fast as the image of the second star. Individual stars show a greater discordance, but the entire series fails to indicate any appreciable systematic variation. The evidence is not conclusive, on account of the small number of stars and the color of some of them. A much better test is afforded by Plate 245. This photograph was taken on January 5, 1886. Two exposures were made of the vicinity of the north pole. The first was made with the clock attached, but the axis was so far out of adjustment that the resulting elongation of the images caused them to appear as trails of a considerable length. The second exposure was made without clockwork. All the trails during the first exposure had the same length, whatever the position of the stars. The trails formed during the second exposure had a length proportional to their polar distance. Accordingly, for stars near the pole the second trail was much the most intense, while the opposite effect was produced for the more distant stars. A discussion of the relative intensity of the trails of all these stars affords an additional determination of the light ratio corresponding to one magnitude. Plate 238, taken on December 30, 1885, was made in a similar way. Another determination might be made by varying the velocity of the images. This could be done by changing the rate of the clock, or the position of the polar axis.

The above methods depend on observations of trails. Similar processes may be applied to the images formed with clockwork, when the instrument and the clock are accurately adjusted. A scale for comparison may be made by taking a series of photographs with different apertures, or varying the time of exposure, moving the telescope a little in right ascension or declination between successive exposures. Two such exposures may be made on the plate to be measured, and the images compared by the method employed by Argelander for observing variable stars. Each image will thus be compared with others of nearly equal brightness, and the final values in grades may be reduced to ratios of light by the relative times or apertures employed in the two exposures. Plate 209, described on page 210, may be reduced in this way. Two exposures were made, of one and five minutes respectively, so that the brighter image of each star represents five times the intensity of the fainter image. Plate 375, taken on February 22, 1886, can be used in the same way. It represents the Nebula of Orion, with exposures of 16^m, 8^m, 4^m, 2^m, 1^m, $30'$, and $15'$. Besides measuring the relative brightness of the stars, this plate permits the relative intensity of different portions of the nebula to be measured. Similar results were obtained on Plates 368, 372, 374, and 382.

The large angular aperture of the lens employed is especially advantageous in photographing the fainter portions of this nebula. An exposure of about 5^m gives the best image of the central portions of the nebula. With 15^m the central portion is completely burned out, and the nebula has as great an extent as is shown in the beautiful photograph obtained by Mr. Common on January 30, 1883. This photograph was taken with the 36-inch reflector, and had an exposure of 30 minutes. The images of the stars in Mr. Common's photograph, owing to the greater focal length of the telescope, are much smaller and better defined than in the Cambridge photographs. The latter, with an exposure of an hour, gives an extension to the nebula of about a square degree. The nebulosity around c *Orionis* is shown with much detail in these photographs.

The nebulosity around the star Maia in the Pleiades, discovered by MM. Henry, was confirmed by Plate 104, taken on November 3, 1885. Although this nebulosity was at once recognized, it was ascribed to a defect in the plate until the true explanation was given by MM. Henry. The first photograph of this object by these gentlemen was taken thirteen days later, or on November 16, 1885.

Jupiter's satellites, as might be expected from their brightness, gave well-marked trails when the telescope was at rest. Some experiments were accordingly made to determine whether the times of the eclipses of the satellites could be advantageously observed by photography. Were it not for the presence of the planet, this would probably be the most accurate method of determining these times, employing the method described on page 211. When the telescope was driven by clockwork, good images of the satellites were obtained in two seconds, and the images were overexposed when the exposure much exceeded ten seconds. During an eclipse the experiment was tried of making a series of exposures, each lasting for ten seconds, at regular intervals of fifteen seconds. The telescope was moved a short distance in right ascension or declination after each exposure, and during the five seconds preceding the usual exposure. Excellent images of the more remote satellites were obtained, but, owing to the short focal length of the telescope, the image of the satellite undergoing eclipse was obscured by that of Jupiter, which was large, on account of the length of the exposure.

Conclusions.

The work in stellar photography done at the Harvard College Observatory may be summarized as follows. The first stellar photograph ever taken was obtained here in 1850. In 1857 the investigation was resumed, and the value of stellar photography as a means of determining the positions and brightness of the components of double stars was established. In 1882, the present research was undertaken with a lens having an aperture of only 2¼ inches. It was shown that photography could be used as a means of forming charts of large portions of the sky, and of determining the light and color of stars in all portions of the heavens. Photographs of the trails of close polar stars no brighter than the eleventh magnitude were obtained without clockwork. Stellar spectra were obtained of the brighter stars without clockwork, in which all the principal lines were well shown. In 1885 the investigation was resumed with a telescope having an aperture of 8 inches. With this, 117 stars within one degree of the pole, one of them no brighter than the fourteenth magnitude, left trails. The average deviation of the measures of the brightness of these stars on different photographs was less than a tenth of a magnitude, a greater accordance than is given by any other photographic method. A similar result was obtained from the Pleiades, of which group over fifty left trails. Similar trails are now being obtained of the stars north of −30° in all right ascensions. This work began in the autumn of 1885 at 23ʰ, and has already been completed for more than half of the sky. By photographing on the same plate polar stars near their upper and lower culminations, material has been accumulated for determining the atmospheric absorption on each night of observation. A study has been made of the application of photography to the transit instrument. Measurements of the trails show that the position of a star may be determined from its trail with an average deviation of 0′.03, which is about one half the corresponding deviation of eye observations.

Charts may be constructed 5° square, having the same scale and dimensions as those of Peters and Chacornac. A single exposure of one hour is required, and it is not necessary that the observer should remain with his eye at the telescope to correct the errors of the clock.

By placing a large prism in front of the object-glass, excellent stellar spectra have been obtained. An exposure of five minutes gives the spectra of all stars brighter than the sixth magnitude in a region 10° square. About half of the region north of −25°, beginning at 0ʰ 0ᵐ, has been photographed in this way. With an exposure of an hour the spectra of stars no brighter than the ninth magnitude are shown. Over a hundred stars have thus been taken simultaneously on a plate by a single exposure. Means have been provided for carrying out this work on an extended scale, as a memorial to the late Dr. Henry Draper.

Miscellaneous observations have been secured of the Pleiades, of the Nebula in Orion, of Jupiter's satellites, and of various other objects; also of the new star in Orion and of its spectrum, and one plate showing that this star must have been much fainter on November 9, 1885, than when discovered, five weeks later.

www.ingramcontent.com/pod-product-compliance
Lightning Source LLC
Chambersburg PA
CBHW031806090426
42739CB00008B/1188